Springer Tracts in Modern Physics 117

W0235014

Editor: G. Höhler
Associate Editor: E. A. Niekisch

Editorial Board: S. Flügge H. Haken J. Hamilton
W. Paul J. Treusch

Springer Tracts in Modern Physics

---

* denotes a volume which contains a Classified Index starting from Volume 36

P.K. Khabibullaev
B.G. Skorodumov

Determination of Hydrogen in Materials
Nuclear Physics Methods

With a Foreword by
L.G. Earwaker and G.H. Sicking

With 38 Figures

Springer-Verlag Berlin Heidelberg GmbH

Professor Dr. Pulat K. Khabibullaev
Dr. Boris G. Skorodumov

Institute of Nuclear Physics, Uzbek SSR Academy of Sciences
702132 Tashkent, USSR

Manuscripts for publication should be addressed to:
Gerhard Höhler
Institut für Theoretische Kernphysik der Universität Karlsruhe
Postfach 6980, D-7500 Karlsruhe 1, Fed. Rep. of Germany

*Proofs and all correspondence concerning papers in the process of publication
should be addressed to:*
Ernst A. Niekisch
Haubourdinstraße 6, D-5170 Jülich 1, Fed. Rep. of Germany

Library of Congress Cataloging-in-Publication Data. Khabibullaev, P. K. (Pulat Kirgizbaevich). Determination of hydrogen in materials: nuclear physics methods / P.K. Khabibullaev, B.G. Skorodumov; with a foreword by L.G. Earwaker and G.H. Sicking. p. cm. – (Springer tracts in modern physics; 117) Bibliography: p. Includes index. 1. Materials-Hydrogen content-Measurement. 2. Nuclear physics-Technique. I. Skorodumov, B.G. (Boris G.), 1934–. II. Title. III. Series. QC1.S797 vol. 117   [TA410]   539 s-dc20   [620.1′12]   89-11332

ISBN 978-3-662-15087-0   ISBN 978-3-540-46067-1 (eBook)
DOI 10.1007/978-3-540-46067-1
© Springer-Verlag Berlin Heidelberg 1989
Originally published by Springer-Verlag Berlin Heidelberg New York in 1989.
Softcover reprint of the hardcover 1st edition 1989

2157/3150-543210 – Printed on acid-free paper

Preface

Hydrogen is the most prevalent element of the cosmos, appearing in both organic and inorganic matter. Although among all the elements the structure of hydrogen is the most simple, its detection and chemical analysis (qualitative and quantitative) is anything but simple.

Most nuclear methods for detecting, analysing and depth profiling hydrogen in near-surface regions of materials were suggested and tested in the 1970s. This decade is likewise marked by a rapidly increasing scientific interest in metal-hydrogen systems. These developments are, of course, not unrelated.

In fact it was the energy crisis of 1973–74 that caused an upward swing in research activities on metal-hydrogen systems. It is their potential to solve future energy problems which, even more than ten years later, makes hydrogen in materials a fascinating subject of research.

There are many experimental methods by which metal-hydrogen systems can be studied. However, among these there are only a few, namely the ion beam analysis methods, which are able to directly probe the hydrogen and to measure its distribution in near-surface regions of materials. In this monograph nuclear methods for the determination of the concentration of hydrogen and its isotopes in materials, in particular ion beam analysis, are reviewed. The methods are based on the bombardment of materials with accelerated ions, γ rays, or neutrons and subsequent analysis of the products of the induced nuclear reactions.

This book is intended to be an introduction for readers interested in applying these techniques to their own measurement problems.

Tashkent, August 1989

P.K. Khabibullaev
B.G. Skorodumov

Contents

List of Symbols, Variables and Abbreviations

$A(a,b)B$	presentation of nuclear reaction: A — initial nucleus, a — incident particle, b — emitted particle, B — residual nucleus (p. 21)
m	particle mass (p. 10)
v	particle velocity (p. 20)
z	particle atomic charge (atomic number) (p. 13)
Z	atomic number of substance (p. 20)
A	atomic weight (p. 20)
n_0	atomic density (p. 11)
C	concentration (p. 11)
E	particle kinetic energy (p. 9)
ΔE	energy loss (p. 9); energy resolution (p. 19)
ψ	angle between beam direction and sample surface (p. 10)
θ	angle between beam and emitted particle directions (p. 10)
$\Delta\Omega$	solid angle of detection (p. 10)
S	stopping power (p. 9)
$\tilde{S}$	stopping parameter (p. 19)
s	standard deviation (p. 17)
R	particle range (p. 9); resonance designate (p. 13)
σ	cross section (p. 10)
$d\sigma/d\Omega$	differential cross section (p. 10)
Q	nuclear reaction energy (p. 10)
x	depth of an impurity location in the sample (p. 11)
Δx	depth resolution (p. 18)
l	thickness of a sample (p. 10)
N	number of particles (p. 10)
η	detection efficiency (p. 11)
α	α particle (p. 9); constant in eq. $R(E) = \alpha E^\beta$ (p. 13); angle (p. 22); constant in (6.3) (p. 72); α phase (p. 30); reliability coefficient (p. 15)
β	constant in eq. $R(E) = \alpha E^\beta$ (p. 13); β phase (p. 30)
k	kinematic factor (p. 13)
γ	γ rays (p. 9); energy width of a channel (p. 60)
K	calibration factor (p. 15); factor in (3.12) (p. 20); channel number (p. 58)
Γ	resonance width (p. 14)

P	probability (p. 17)
ε	relative error (p. 17)
a	factor in (3.11) (p. 20); constant in (6.6) (p. 79)
$\bar{\vartheta}^2$	mean-square angle (p. 20)
I	relationship of monitor readings (p. 12); intensity (p. 29)
D	diffusion coefficient (p. 69)
t	time (p. 69)
τ	reduced time (p. 71); exposure time (p. 76)
T	temperature (p. 73)
at.H/at.M	number of H atoms per material atom (p. 11)
at.ppm	number of atoms per million (p. 11)
at.%	atomic per cent (p. 27)
wt.%	weight per cent (p. 29)
fwhm	full width at half maximum (p. 18)
IBA	ion beam analysis (p. 8)
N(R)RA	nuclear (resonance) reaction analysis (p. 32)
ERD(A)	elastic recoil detection (analysis) (p. 52)
PIXE	proton induced x ray emission (p. 56)
RBS	Rutherford backscattering spectrometry (p. 52)
TPHC	time-to-pulse height converter (p. 58)
UHV	ultra high vacuum (p. 33)

Introductory Preface

By L.G. Earwaker and G.H. Sicking

With 1 Figure

Hydrogen is the most prevalent element of the cosmos. It is met within inorganic, organic and biologic matters alike. There are three isotopes, $^1_1$H, $^2_1$H and $^3_1$H, having atomic masses of 1.00797, 2.01400 and 3.01605, respectively. Since the relative mass differences are the biggest encountered throughout the periodic table, chemical isotope effects are well pronounced. Consequently the hydrogen isotopes are distinguished by extra-names and extra-symbols: protium H ($^1_1$H), deuterium D ($^2_1$H) and tritium T ($^3_1$H). The positive ions are called proton, deuteron and triton. With respect to particle beams their symbols are p, d and t. The natural abundance of protium is 99.985% and of deuterium 0.015%. Tritium is unstable and decays by emission of β particles, having a maximum energy of 18.61 keV: $^3_1$H $\rightarrow$ $^3_2$He $+ _{-1}^{0}e + \tilde{v}$; $\tau = 12.26$ years. Among all the elements the structure of the hydrogen atom is most simple but its detection and chemical analysis in materials (qualitatively and quantitatively) is anything but simple.

Most of the nuclear physics methods for detecting, analysing and depth profiling hydrogen in near-surface regions of materials were suggested and put to the test in the 1970s. This decade is like-wise marked by a rapidly increasing scientific and technological interest in metal/hydrogen systems. The development of these trends were, of course, not independent of each other.

In fact it was the energy crisis around the turn of the year 1973/74 that caused an enormous upward swing in the research activities on technical applications of metal/hydrogen systems. This was based mainly on some basic studies on hydrogen in intermetallic compounds in the late 60s, leading to the concept of hydrogen in metals as a mobile energy source for automotive purposes [1–6]. In the following years the research activities were supported generously and worldwide by national science foundations, governments and industries. At least two series of International Conferences, one on "Hydrogen in Metals", the other on "The Properties and Applications of Metal-Hydrides" shot up.

The first series developed from an early and inspiring meeting of only a handful of scientists at the University of Münster, FRG, in 1968. The second was started by a symposium held in Geilo, Norway, in the late summer 1977. Both were merged together to one series of biennial conferences in 1988.

The first particle accelerator was built by Cockcroft and Walton in 1932, but it took more than 30 years before accelerators were commonly used in material analysis. This late application as an analytical tool is the result of a number of circumstances. Originally particle accelerators were built for breaking nuclei into smaller particles. In 1932 merely proton, electron and neutron were known as

1

constituent parts of matter. The first generation of accelerators helped to find some more particles, as e. g. neutrino, positron, muon and pion.

Nuclear science soon required bigger accelerators with more than a few MeV beam energy, which is about the binding energy of a nucleon. The first proton-accelerator with more than 1 GeV beam energy was finished at Brookhaven, USA, in 1952. An irritatingly large number of particles, which could not be believed to be elementary particles, have been discovered since. The question, which particles are elementary and by which kind of interaction they stick together, called for bigger accelerators and the Tevatron at Fermilab, USA, is the latest land-mark, being designed for 1 TeV beam energy.

Accelerators with 5 TeV (Pentevac at Fermilab, USA) and 3 TeV (UNK at Serpuchow, USSR) are planned or under construction.

The way followed by particle physics entailed free capacities at particle accelerators with beam energies up to about 30 MeV. Encouraged by the rapid growth of semi-conductor technologies these accelerators were applied to research fields outside basic nuclear physics. This happened in the 1960s. Backscattering Spectrometry [7] became an important analytical tool for monitoring near-surface regions of semi-conductors. Ion implantation as a preparation technique supplemented the analytical techniques, whereby both could be carried out within the same accelerator.

New accelerators were built for these purposes in the 1960s and 70s and two international series of conferences developed in this field. The first, "The Application of Accelerators in Research and Industry" was started in Oak Ridge, TN, USA, in 1968. The beginning of the second series "Ion Beam Analysis (IBA)" was in 1973 in Yorktown Heights, NY, USA.

Both these symposia regularly bring together scientists from most areas of natural sciences. This is due to the fact that non-destructive material analysis is a central problem in many research fields and likewise in many domains of industrial production. Medicine and biology are as well concerned as geochemistry, metallurgy, material sciences and, of course, chemistry and physics. Construction materials, refractory metals, semi-conductors, thin films, glassy materials are all industrial products investigated by ion beam analysis. Corresponding phenomena are corrosion, hardening, diffusion, impurity distribution, catalysis, coatings, interfacial mixing, and passivation. The catalogue of matters and features being subject to ion beam analysis may be lengthened and it is not surprising that a spill-over to metal/hydrogen-systems happened. The research field of hydrogen in metals touches upon many classical problems, as e. g. chemical bonding, electronic structure, lattice structure, magnetism, phase transformation, isotope effects, superconductivity, reaction kinetics, thermodynamics, diffusion, permeation, segregation, decomposition, lattice defects, critical point phenomena, surface chemistry, chemisorption, catalysis, electrochemistry, purification processes, material embrittlement, heat pumps, heat exchangers, fuel cells, batteries, lattice dynamics, etc.....

The experimental tools for investigating these problems may casually be arranged into two categories:

i) The first group contains experiments, which probe the properties of the metal/hydrogen system as a whole. Absorption-isotherms and band-magnetism are good examples for these experiments.

ii) The second group contains experiments, which probe the hydrogen itself, stored in the system. Neutron scattering and NMR (nuclear magnetic resonance) may be taken as examples in this group.

There are, of course, a number of experimental methods, which do not fit in one of these categories in an unconstrained way. Internal friction experiments may be mentioned in this context. Experimental methods belonging clearly to group (ii) are not very numerous. As this book proves, nuclear physics methods stand out among them for several reasons. The most important is the capability of these methods to measure concentration profiles of hydrogen in materials. It is a surprising fact that up to the end of 1973 only a few papers deal with hydrogen profiling [8 –11]. *Cohen* et al. [8] employed the proton-proton scattering technique for monitoring the hydrogen concentration profile across a 67 μm thick pile of four Mylar-foils alternating with three Fe-foils. The authors demonstrated the detection limit of this method to be 1 ppm H. In 1973 *Leich* et al. [10,11] established the resonant nuclear reaction $^1H(^{19}F,\alpha\gamma)^{16}O$ for measuring hydrogen depth profiles in solids (lunar samples).

Hydrogen depth profiling made considerable advances with the introduction of the ^{15}N-method [12-14] in 1976. This nuclear resonance reaction $^1H(^{15}N, \alpha\gamma)^{12}C$ (see Sect. 5.1a and g of the booklet) offers outstanding depth resolution, due to the small resonance width of the reaction, very good detection sensitivity, and a fairly good analysable depth of about one micrometer.

Quite a number of publications dealing with the ^{15}N-method are given in this booklet and some more references may be added [15–24]. Two of the papers show H-concentration curves vs. depth after different heat-treatments, which then are evaluated with respect to the diffusivity of H in Ti and a TiV-alloy [15] and in Ta [17]. Other papers deal with trapping of hydrogen at the location of Ti-atoms implanted in Fe [18, 22] and Ni [18] or with trapping of hydrogen in the Al/SiO_2 interface of Al-gate MOS-structures [24]. Similarly in [23] hydrogen was found to enrich just below oxide layers on Zr and Zr-2.5 wt%Nb alloys. Pd-atoms implanted in Ti near-surface layers were found [19] to improve the kinetics of hydrogen uptake. In [20] H-depth profiles in films of amorphous CuZr-alloys were reported and in [16] the reverse of the ^{15}N-method, $^{15}N(^1H, \alpha\gamma)^{12}C$ was used for investigating the nitrogen refining of stainless steel.

The ^{15}N-method (and other resonance reactions) is strictly specific to the particular hydrogen isotope, whereas the method of elastic recoil detection (ERD or ERDA) allows depth profiling of all the three hydrogen isotopes and of other low Z elements. This method was first employed in the late 70s [25–30] although a special case of ERD (proton-proton scattering, see Sect. 5.3c) was already applied in 1972 [31].

This book distinguishes between ERD with heavy ion beams as e. g. ^{35}Cl-ions (Sect. 5.3a) and ERD with a He-beam (Sect. 5.3b). The principle of ERD is to bombard the target under a small incidence angle of typically 20° and to look for the hydrogen atoms scattered out of the target in a forward direction. With this geometry, the primaries scattered into the same direction have to be shielded from the detector system. For this purpose a stopper-foil, about $10\,\mu$m of Al, is mounted in front of the detector system, giving rise to an additional energy straggling of the recoiled hydrogen particles. This, together with multiple scattering effects within the target, is the main reason for a reduced depth resolution of merely about 50 nm (2 MeV He-beam), about one order of magnitude worse than is possible with the ^{15}N-method. For further details see Sect. 5.3 of this book and [32]. Some more references of work done by ERD are [33–41].

The stopper-foil, the main source of a reduced depth resolution, can be successfully replaced by crossed electric and magnetic fields ($E \times B$ filter) as is shown in [34]. These authors report a depth resolution of 5 nm–20 nm, depending on depth. Since they used a primary ^{4}He-beam of only 350 keV energy the probing depth is limited to about 100 nm. The $E \times B$ filter technique is used by [39] for measuring the concentration profile of H implanted into TiC-coatings.

Gas telescope detectors [36] and solid state coincidence telescope detectors [38] are frequently used (especially in conjunction with heavy ion beams or He-beams of more than about 20 MeV) for simultaneously recording all kinds of recoiled particles and to identify them by their mass. This is reported for elements from H to O in the periodic table [36]. In [38] the simultaneous detection of H, D and T is described. These authors also report the quantitative analysis of a Zr-hydride, containing protium and deuterium as well. From this kind of measurement the equilibrium distribution of hydrogen isotopes between gas-phase and solid-phase can be evaluated [38]. The diffusion of Ti into $LiNbO_3$ (Ti : $LiNbO_3$ diffused wave guides) was studied by [35]. These authors used ERD with 30 MeV Cl-ions for recoiling H (and O), originating from the preparation conditions. Some glass-samples with simulated nuclear damage, leached in water, were depth-profiled with respect to H using ERD with 20 MeV Si^{4+}-ions [33].

The surfaces of Be-plates and Ni-films were investigated for the number of adsorbed H-atoms by applying ERD with a 50 MeV Ar-beam [41]. These investigations concern the problem of finding suitable materials for neutron bottles. These materials have to be mostly free of adsorbed hydrogen in order to preserve ultracold neutrons over long times. ERD with He-ions of 2 MeV energy is used for determining H-concentration profiles in natural topaz stone [37]. In [40] the profile of H implanted into Ti was measured (1.9 MeV ERD with He) after annealing in order to determine redistribution and trapping mechanisms. H implantation at low temperatures was found to produce TiH_2 precipitations.

2.8 MeV ERD with He ions was employed [38] for investigating the surface-activation process of Laves-phase hydrogen storage materials under in-situ treatment with hydrogen (see Fig. 1).

Since the early days of *Thomas Graham* [42] metal/hydrogen-systems have attracted many scientists and still do so.

This fascination may be traced back to the fact that hydrogen in metals quite often shows up an unexpected behaviour. The superconductivity of PdH_n with $n > 0.75$ [43] belongs to this kind of phenomena, and the recent preliminary note [44] that hydrogen isotopes in palladium electrodes may undergo a cold

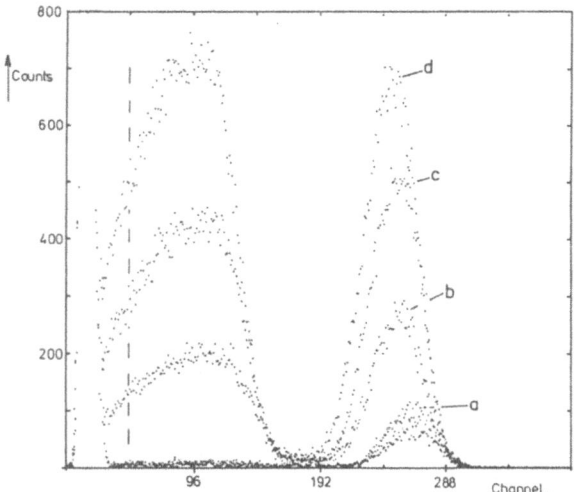

Fig. 1. The figure shows a juxtaposition of the ERD-spectra obtained from an off-stoichiometric TiMn-Laves-phase $(Ti_{0.4}Mn_{0.6})$: (a) before hydrogen treatment (b) after exposure to hydrogen (15 mbar, 10 minutes) at $200°$ C (c) after another exposure (20 mbar, 30 minutes) at $200°$ C (d) after a final treatment (30 mbar, 30 minutes) at $200°$ C. A surface hydride is clearly seen to develop, separated by a hydrogen depletion zone from the likewise growing hydrogen concentration in the bulk. With increasing hydrogen treatment the depletion zone shrinks at the expense of the surface hydride

fusion is another example. Precisely this proximity to the future energy problems of mankind makes hydrogen in materials a fascinating subject of research. There are many experimental methods by which metal/hydrogen-systems can be studied. However among these there are only a few, namely the ion beam analysis methods, which are able to directly probe the hydrogen and to measure the distribution of hydrogen in near-surface regions of materials.

This book may be taken as an introduction to ion beam analysis and it fulfils its purpose if some of its readers are encouraged to apply these methods to their particular problems.

References

1 J.J. Reilly, R.H. Wiswall: Inorg. Chem. **6**, 2220 (1967)
2 J.J. Reilly, R.H. Wiswall: Inorg. Chem. **7**, 2254 (1968)
3 K. Hoffmann, W.E. Winsche, R.H. Wiswall, J.J. Reilly, T.V. Sheehan, C.H. Waide: "Metal Hydrides as a Source of Fuel for Vehicular Propulsion", Int. Automotive Eng. Congress, Detroit, S.A.E. 690, 232 (1969)
4 J.H.H. van Vucht, F.A. Kuijpers, H.C.A.M. Bruning: Philips Res. Rep. **25**, 133 (1970)
5 R.H. Wiswall, J.J. Reilly: "Method of Storing Hydrogen", U.S. Patent, 3, 516, 263 (June 23, 1970)
6 F.A. Kuijpers, H.H. van Mal: J. Less-Common Met. **23**, 395 (1971)
7 Wei–Kan Chu, J.W. Mayer, M.A. Nicolet: "Backscattering Spectrometry" (Academic Press, New York, San Francisco, London 1978)
8 B.L. Cohen, C.L. Fink, J.H. Degnan: J. Appl. Phys. **43**, 19 (1972)

9   R.S. Blewer: Appl. Phys. Lett. **23**, 593 (1973)

10 D.A. Leich, T.A. Tombrello: Nucl. Instrum. Methods **108**, 67 (1973)

11 D.A. Leich, T.A. Tombrello, D.S. Burnett: Geochim. Cosmochim. Acta **2**, 1597 (1973)

12 W.A. Lanford, H.P. Trautvetter, J.F. Ziegler, J. Keller: Appl. Phys. Lett. **28**, 566 (1976)

13 W.A. Lanford: Science **196**, 975 (1977)

14 W.A. Lanford: Nucl. Instrum. Meth. **149**, 1 (1978)

15 E. Brauer, R. Doerr, R. Gruner, F. Rauch: Corrosion Science **21**, 449 (1981)

16 J.L. Whitton, M.M. Ferguson, G.T. Ewan, I.V. Mitchell, H.H. Plattner: Appl. Phys. Lett. **41**, 150 (1982)

17 M. Weiser, S. Kalbitzer, M. Zinke–Allmang, H. Damjantschitsch, G. Frech: Materials Science and Engineering **69**, 411 (1985)

18 H. Baumann, Th. Lenz, F. Rauch: Materials Science and Engineering **69**, 421 (1985)

19 B. Hoffmann, H. Baumann, F. Rauch: Nucl. Instrum. Methods, **B15**, 361 (1986)

20 M. Fallavier, M.Y. Chartoire, J.P. Thomas: Nucl. Instrum. Methods, **B15**, 712 (1986)

21 M. Fallavier, J.P. Thomas, J.M. Frigerio, J. Rivory: Solid State Commun. **57**, 59 (1986)

22 B. Hoffmann, H. Baumann, F. Rauch, K. Bethge: Nucl. Instrum. Methods, **B28**, 336 (1987)

23 A. Stern, D. Khatamian, T. Laursen, G.C. Weatherly, J.M. Perz: J. Nucl. Materials, **148**, 257 (1987)

24 A.D. Marwick, D.R. Young: J. Appl. Phys. **63**, 2291 (1988)

25 I.P. Chernov, J.P. Matusevich, V.P. Kosyr: Atomnaja Energija **41**, 51 (1976)

26 J. L'Ecuver, C. Brassard, C. Cardinal, J. Chabbal, L. Dèschenes, J.P. Labrie, B. Terrault, J.G. Martel, R.St. Jacques: J. Appl. Phys. **47**, 881 (1976)

27 B. Terrault, J.G. Martel, R.G.St. Jacques, J. L'Ecuver: J. Vac. Sci. Technol. **14**, 492 (1977)

28 J. L'Ecuver, C. Brassard, C. Cardinal: Nucl. Instrum. Methods **149**, 271 (1978)

29 B.L. Doyle, P.S. Percy: "The Analysis of Hydrogen in Solids", US Department of Energy, DOE/ER–0026, 92 (1979)

30 B.L. Doyle, P.S. Percy: Appl. Phys. Lett. **34**, 811 (1979)

31 B.L. Cohen, C.L. Fink, J.H. Degnan: J. Appl. Phys. **43**, 19 (1972)

32 A. Turos, O. Meyer: Nucl. Instrum. Methods **B4**, 92 (1984)

33 G.W. Arnold, J.C. Petit: Nucl. Instrum. Methods **209/210**, 1071 (1983)

34 G.G. Ross, B. Terreault, G. Gobeil, G. Abel, C. Boucher, G. Veilleux: J. Nucl. Materials **128/129**, 730 (1984)

35 P.M. Read, S.P. Speakman, M.D. Hudson, L. Considine: Nucl. Instrum. Methods **B15**, 398 (1986)

36 A.M. Behrooz, R.L. Headrick, L.E. Seiberling, R.W. Zurmühle: Nucl. Instrum. Methods **B28**, 108 (1987)

37 M. Rubel, H. Bergsåker, B. Emmoth, S. Nagata: Nucl. Instrum. Methods **B28**, 284 (1987)

38 P.W. Albers, G.H. Sicking, L.G. Earwaker, J.B.A. England: Ber. Bunsenges. Phys. Chem. **91**, 573 (1987)

39 D. Fournier, F.G.St. Jacques, G.G. Ross, B. Terreault: J. Nucl. Materials **145–147**, 379 (1987)

40 Y. Takeuchi, N. Imanishi, K. Toyoda, T. Uchino, M. Iwasaki: J. Appl. Phys. **64**, 2959 (1988)

41 Y. Kawabata, M. Utsuro, S. Hayashi, H. Yoshiki: Nucl. Instrum. Methods **B30**, 557 (1988)

42 T. Graham: Proc. Roy. Soc. (London) **156**, 399 (1866)

43 F. Skoskiewicz: Phys. Status Solidi (a) **11**, K 123 (1972)

44 M. Fleischmann, S. Pons: J. Electroanal. Chem. **261**, 301 (1989)

Determination of Hydrogen in Materials
Nuclear Physics Methods

By P.K. Khabibullaev and B.G. Skorodumov

With 37 Figures

1. Introduction

Hydrogen plays a key role in many scientific and technical problems, and the hydrogen-metal system (to be more exact, hydrogen-material system) has emerged as an area of special interest in modern solid state physics. The current state of knowledge of the hydrogen-metal system is reviewed in [1] and also reflected in the Proceedings of some International Symposia [2, 3]. In spite of the fact that investigations have been in progress for many years, the theory of the interaction of hydrogen with other materials is far from being clear, and the emergence of new experimental techniques has resulted in new problems to be solved.

The presence of hydrogen may have important effects on physical, chemical and electrical properties of many materials. The negative influence of hydrogen on properties of metals is exemplified by hydrogen embrittlement that causes great damage in industry [4]. On the other hand, the presence of hydrogen in some materials may improve their working properties [5, 6]. Thus, hydrogen may serve as an important tool for creating materials with specific characteristics, and investigation of material-hydrogen systems has become more urgent due to the necessity of solving several important problems in the study of energy-related materials.

Fundamental studies of hydrogen behaviour in materials must be undertaken in order to solve scientific and applied problems related to the hydrogen-metal system. In turn, this requires development of new analytic techniques that differ from the traditional ones in terms of high accuracy, selectivity and sensitivity. These techniques should not only provide the information about the total hydrogen content, but also about its depth distribution in the materials studied.

Hydrogen is the most difficult atomic species to analyse with many traditional methods. Because of its low Z, methods based on x ray or Auger emission do not work, and, because of its light mass, Rutherford backscattering can not be used. A hydrogen depth distribution can be directly measured with high sensitivity by the powerful method of secondary ion mass spectrometry (SIMS). However, it is naturally destructive and suffers from lack of discrimination between different concentrations. Besides, radiation damage introduced during the SIMS analysis seem to change the lattice structure at the surface. These disadvantages are therefore responsible for the restricted usefulness of SIMS for studying the behaviour of the metal-hydrogen system [7].

7

Nuclear physics methods of non-destructive elemental analysis of materials, including the determination of hydrogen and its isotopes, are now being intensively developed. These methods are based on the bombardment of materials by accelerated particles, and the examination of the products of their nuclear interaction with the element analysed. In contrast to nuclear activation analysis, where irradiation of specimens and measurements of their radioactivity take place at different times, the methods mentioned are called instantaneous or rapid nuclear analysis as the measurements are made simultaneously with irradiation. Charged particle participation in a nuclear interaction allows one to perform concentration depth profiling. Various high energy (i.e. in the MeV range) ion-beam techniques are widely used in materials analysis. They are known under the general name of Ion Beam Analysis (IBA) to which an International Conference is devoted every two years.

Hydrogen profiling methods, using ion beams, are reviewed in [8,9] which report investigations up to 1977. Lately, interesting publications have appeared that discuss new application of well-known methods as well as the original developments.

In this review the methods for the determination of concentration of hydrogen and its isotopes in materials, based on the results of the interaction of accelerated ions, γ rays and neutrons with hydrogen dissolved in materials, are discussed using the literature data up to 1987 and investigations carried out by the authors. The material is distributed over 6 chapters: Chap. 1 contains the introduction. The principles of nuclear physics analysis are considered in Chap. 2, where a general relationship between the number of recorded products of given energy, and hydrogen concentration at a definite depth of a sample is obtained. Chap. 3 is devoted to the analytical characteristics of nuclear physics methods, i.e. the detection limit (sensitivity), rapidity, selectivity and accuracy that have general significance, as well as depth resolution and maximum probing depth, that appear to be distinctive features of techniques used for hydrogen profiling. The existing methods are tabulated and certain criteria for choosing a method are given.

In Chap. 4 methods capable of determining the total hydrogen content are described, and Chap. 5 goes on to deal with the specific reactions used for hydrogen depth profiling, both by resonance methods and energy analysis. The latter addresses nuclear reactions proper and elastic scattering. Of these, singled out are the heavy ion elastic recoil technique and methods devised by the authors that are based on elastic scattering of protons and neutrons. Examples of the applications of some of the methods described are also presented. Particular attention is focused on the use of nuclear depth profiling in the direct study of stationary and non-stationary diffusion of hydrogen isotopes. This aspect of the application of nuclear physics methods is considered in Chap. 6.

Information about the lattice location of hydrogen atoms in crystals can be obtained by application of the technique of particle channeling [10]. However, this aspect of the problem is beyond the aims of the present review, as are the investigation of hydrogen-material systems using the perturbed γ–γ angular correlation technique [11], the Mössbauer effect and quasielastic neutron scattering [1] that provide the data on processes on a microscale. We also do not deal with the analysis of tritium using methods based on its radioactivity [12]. The only sources cited are those that most fully summarize the essence of the method and contain all the necessary references.

2. Principles of Nuclear Physics Analysis

Nuclear methods for the determination of hydrogen content are based on its nuclear interaction with incident radiation. Neutrons, γ rays from isotope sources and reactors, accelerated ion beams of various mass and energy, bremsstrahlung radiation and neutrons generated by reactions of accelerated ions with nuclei (neutron generators) are all used. The interaction products, i.e. γ rays, neutrons, and charged particles, are detected by scintillation and semiconductor detectors. Some methods use only the *total number* of particles that appear as a result of nuclear reactions with hydrogen, while others utilize their energy spectrum. The measurement of charged-particle energy may be accompanied by identification of the particle type, which is often performed with the help of a $\Delta E - E$ detector telescope. By measuring the energy loss ΔE and total energy of particle E, the type of a particle may be determined using the relationship $\Delta E \cdot E \approx$ const, as the constants are different for protons, deutons, tritons, α particles, etc.

During irradiation, interaction occurs with the host material as well as with hydrogen contained in it. The possibility of identifying the events corresponding to reaction with hydrogen from among the competing processes is conditioned by the choice of the type and energy of bombarding particles as well as by the technique for the detection of reaction products. The sample material acts as a medium in which incident and emitted radiation loses energy. The processes accompanying the passage of nuclear radiation through matter are described in, for example, [13]. Here we review briefly the basic ideas necessary to understand the physical principles of the method presented.

The type of interaction with matter depends on the nature of the radiation. γ rays interact mainly with the electron shells of atoms by photoelectric absorption, the Compton effect, and pair formation, but sometimes they may cause so-called photonuclear reactions. Neutrons interact only with atomic nuclei, thereby losing energy as a result of elastic and inelastic scattering. While moving through matter, charged particles interact with electron shells of atoms as well as with nuclei. The former process dominates, and in penetrating through a thickness dx, a particle loses energy dE as a result of ionization and excitation of the medium atoms. Particle energy losses per [cm] $S(E)$ (or per [g/cm^2] $S_m(E) = S(E)/\varrho$, where ϱ is the substance density) is called the stopping power of a given material for the given type of particles with energy E; the particle range, $R = \int_E^0 S^{-1} \, dE$. The stopping powers $S(E)$ and ranges $R(E)$ of charged particles in various materials have been tabulated [14, 15].

In the case of elastic scattering a certain portion of the incident particle kinetic energy is transferred to the recoil nucleus, but the total kinetic energy remains the same. Inelastic nuclear interactions can be understood as a scattering

9

process with residual nucleus excitation and also as a nuclear reaction process, i.e. the formation of new particles in the process of interaction. The total kinetic energy of particles is not preserved in inelastic processes.

The relative contribution of various processes of nuclear interaction is characterized by a cross-section. The cross-section $\sigma(E)$ is defined as the probability that a given interaction event will take place when one particle with energy E passes through a target containing one interaction center per [cm^2]. It has the dimension of area and is measured in barns (1 b $= 10^{-24}$ cm^2). The dependence of cross-section on the incident particle energy is called the reaction excitation function. It may be smooth or it may have a resonance structure. The narrow resonances which appear in the excitation function are connected with compound nucleus formation. The incident particle energy that leads to resonance is determined by the excitation energy of the corresponding level of a compound nucleus. The characteristics of this level predetermine the height and the width of the resonance peak.

The differential cross-section $d\sigma/d\Omega$ indicates the probability of such an interaction between the particles when one of them is emitted in a solid angle from Ω to $\Omega + d\Omega$. The interaction with an additional limitation of the emitted particle energy (in the range from E to $E + dE$) is characterized by the double differential cross section $d^2\sigma/dE\,d\Omega$.

As the nature of nuclear forces is not known yet, theoretical expressions for the cross-sections of different nuclear processes (except Coulomb scattering) are based on phenomenological models. The cross-sections used in nuclear analysis are often taken from reference books. It should be added that for any nuclear reaction $m_1 + m_2 \rightarrow m_3 + m_4$ (where m_i are the masses of particles participating in a reaction) one can define an energy balance

$$E_1 + Q = E_3 + E_4 + E_\gamma \ .$$

Here $E_i (i = 1, 2, 3, 4)$ are the kinetic energies of the corresponding nuclei (the energy of a target nucleus $E_2 = 0$ in the laboratory frame); E_γ is the overall energy released through γ ray emission (if the residual nucleus occurs in an excited state); Q is the nuclear reaction energy called the reaction Q value, which may be positive (exothermic reaction) or negative (endothermic reaction), or may be equal to zero (elastic scattering). Both the possibility of a nuclear reaction, and the kinetic energy of the reaction products depend on this value. The Q values are calculated from the tabulated atomic mass excesses of the corresponding nuclei.

Now let us consider the physical principles of hydrogen analysis in materials for the most complex case. Here, charged particles with energy E_0 are incident at an angle ψ to the surface of a sample plate of thickness l. At the same time an emitted particle, also charged, is registered by a detector situated either in the front (transmission geometry) or in the back (reflection geometry) hemisphere (Fig. 2.1).

If N_0 particles are incident on a hydrogen-containing sample, then in a target layer of thickness dx the nuclear reaction events with products emitted at an angle θ in the solid angle $\Delta\Omega$ will be

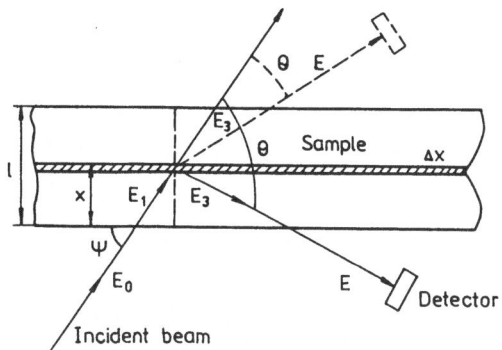

Fig. 2.1. Geometry of hydrogen determination and corresponding symbols; the dotted line corresponds to the transmission geometry

$$dN(x) = \eta(x)N_0\frac{d\sigma}{d\Omega}(E_1(x),\theta)\Delta\Omega n_{0H}(x)\,dx \quad , \tag{2.1}$$

where $E_1(x)$ is the energy of an incident particle just before the reaction at depth x; $n_{0H}(x)$ is the atomic hydrogen density (at.H/cm^3) at depth x; $d\sigma(E_1(x),\theta)/d\Omega$ is the hydrogen differential cross-section; $\eta(x)$ is the detection efficiency (equal to 1 in the case of charged particles, and less than 1 if neutrons or γ rays are detected) which may be dependent on x so far as it depends on detected particle energy E. It is seen from (2.1) that the nuclear reaction yield $(dN(x)/N_0)$ is proportional to the hydrogen content in the layer analysed. The hydrogen concentration $C(x) = n_{0H}/n_{0M}$, i.e. the number of hydrogen atoms taken up per material atom (at.H/at.M) may be obtained from (2.1)

$$C(x) = \frac{dN}{dx}\left[\eta(x)N_0\frac{d\sigma}{d\Omega}(E_1(x),\theta)n_{0M}\Delta\Omega\right]^{-1} . \tag{2.2}$$

Concentration may also be expressed in units of weight, atomic or volume percent, [cm^3] per 100 g of metal, [mg] of hydrogen per 1 g of metal, etc. (1 at.ppm = 10^{-6} at.H/at.M).

If the energy spectrum of reaction products is measured, then, after introducing the channel contents dN/dE (also denoted as $N(E)$) in (2.2), the hydrogen distribution results in

$$C(x) = N(E)\left[\eta(E)N_0\frac{d\sigma}{d\Omega}(E_1,\theta)n_{0M}\Delta\Omega\right]^{-1}\frac{dE}{dx} \quad , \tag{2.3}$$

where the derivative dE/dx (sometimes called a stopping parameter) appears to be one of the main factors that converts the energy spectrum into a concentration profile.

The relative method for the determination of hydrogen content is rather widely used. Comparing the reaction yield obtained from the sample studied with that from the hydrogen standard, the hydrogen concentration can be deduced using the relation

11

$$C(x) = I n_{0Hs} \left[N(E)\frac{dE}{dx}(E) \right]_0 \bigg/ n_{0M} \left[N(E)\frac{dE}{dx}(E) \right]_s , \qquad (2.4)$$

where n_{0Hs} is the hydrogen standard concentration density; $I = (N_{0s}/N_0)$; N_{0s} and N_0 are numbers of incident particles in measurements with standard and sample, respectively. The expressions (2.3,4) determine the hydrogen concentration in a sample layer located at depth x, which can be calculated taking into account the ionization energy losses of incident and emitted particles in the sample material.

An incident particle loses part of its kinetic energy on the way from the surface, where it has the energy E_0, to the collision point with a hydrogen nucleus at depth x, where the interaction takes place with the energy

$$E_1(x) = E_0 - \int_0^{x/\sin\psi} S_1(x')\, dx' . \qquad (2.5)$$

After a reaction an emitted particle of energy E_3 defined by nuclear reaction kinematics also loses energy along its path towards the sample surface where it has the energy

$$E = E_3(E_1, m_i, \theta, Q) - \int_0^{x/\sin(\theta-\psi)} S_2(x')\, dx' \qquad (2.6)$$

in the case of reflection geometry, or

$$E = E_3(E_1, m_i, \theta, Q) - \int_0^{(l-x)/\sin(\psi-\theta)} S_2(x')\, dx' \qquad (2.7)$$

in the case of transmission geometry. Here S_1 and S_2 are the stopping powers of the sample material for incident and emitted particles, respectively. E_3 may be calculated in the non-relativistic case using the relationship

$$E_3 = m_1 m_3 E_1 (\cos\theta \pm p^{1/2})^2 / M , \text{ where} \qquad (2.8)$$

$$M = m_1 + m_2 \text{ and}$$

$$p = \cos^2\theta + M\left[(m_4 - m_1)/m_1 m_3 + m_4 Q/m_1 m_3 E_1\right] .$$

Endothermic reactions are characterized by the energy threshold

$$E_{1th} = M|Q|/m_2 . \qquad (2.9)$$

Below this value the reaction is impossible according to the conservation of energy. To obtain an expression for the energy of the other particle, E_4, it is necessary only to exchange indices 3 and 4 in (2.8).

12

At depths corresponding to small losses of particle energy, the average stopping powers $\overline{S}_1$ and $\overline{S}_2$ are used to find the relationship between E and x. Then the expressions (2.5–7) may be written in a simpler form

$$E_1 = E_0 - \overline{S}_1 x / \sin \psi \ , \tag{2.5'}$$

$$E = E_3 - \overline{S}_2 x / \sin(\theta - \psi) \ , \tag{2.6'}$$

$$E = E_3 - \overline{S}_2 (l - x) / \sin(\psi - \theta) \ . \tag{2.7'}$$

At greater depths, where changes of S are not very small, instead of stopping powers one may use the particle ranges R. As an example, let us find the depth-energy relationship, when a recoil proton is ejected towards the front hemisphere by an elastic collision with an accelerated ion A

$$A + H \rightarrow p + A \ .$$

Using the obvious relationship between the range $R(E)$ and the thickness, x, of a material layer (where a particle loses some energy ΔE as it traverses this layer)

$$R(E) = x + R(E - \Delta E) \ , \tag{2.10}$$

we may write the following system of equations (see Fig. 2.1)

$$\begin{aligned} x / \sin \psi &= R_A(E_0) - R_A(E_1) \\ (l - x) / \sin(\psi - \theta) &= R_p(E_3) - R_p(E) \ . \end{aligned} \tag{2.11}$$

This system may be solved for x, keeping in mind that $E_3 = kE_1$, where the constant k is a kinematic factor; a particle range may be approximated by an expression $R(E) = \alpha E^\beta$; and a proton range

$$R_p(E) \approx z_A^2 R_A(m_A E)/m_A \ , \tag{2.12}$$

where z_A, R_A and m_A are ion charge, range and mass, respectively [13]. This leads to

$$x = \frac{k' R_A(E_0) + l / \sin(\theta - \psi) - R_p(E)}{k' / \sin \psi + \sin^{-1}(\theta - \psi)} \ , \tag{2.13}$$

where $k' = z_A^2 k^\beta m_A^{\beta - 1}$ (β is related to ion A). The ranges $R(E)$, as well as constants α and β, may be found using range-energy curves [15].

Thus, if the energy spectrum $N(E)$ for the products of the nuclear reaction of an incident particle with hydrogen is measured, then each E value of this spectrum may be related to the hydrogen depth x by (2.6,7 or 13). Hydrogen concentration at each particular point may be calculated using (2.3 or 4). This method is known as the energy analysis method.

13

Now let us consider another depth profiling technique, the so-called resonance method. When the reaction excitation function has a strong narrow isolated resonance, with a cross section $\sigma(E_R)$ several orders of magnitude larger than the value outside it, the result is that practically all the interactions with hydrogen occur at depth x_R related to the incident beam energy $E_1(x_R) = E_R$ (where E_R is the resonance energy) (Fig. 2.2). When the energy of accelerated ions is raised to the resonance energy, hydrogen at the surface of the sample will be detected. If the incident beam has an energy above E_R, it must lose some energy before it reaches E_R, and the nuclear reaction occurs at a certain depth x_R inside the sample. Using the expression (2.5'), which is true for the small energy-loss approximation, we obtain

$$x_R = (E_0 - E_R)\sin\psi/\overline{S}_1 \ . \tag{2.14}$$

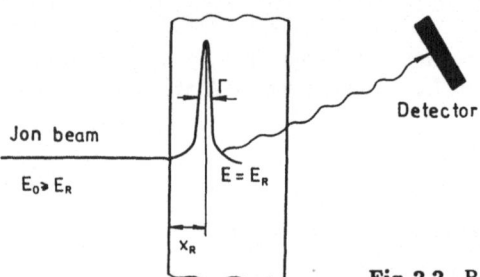

Fig. 2.2. Resonance technique for hydrogen profiling

In such measurements the incident beam energy E_0 is changed in small increments, thereby changing the depth at which the resonance occurs. If the resonance has the width Γ, then the expression for the total number of counts $N(E_0)$ at the accelerator energy setting E may be written

$$N(E_0) \approx N_0 \Delta\Omega\overline{\eta} \int_{E_R+\Gamma}^{E_R-\Gamma} \frac{d\sigma}{d\Omega}(E_1)n_{0H}(x(E_1))S_1^{-1}(E_1)\,dE_1 \ , \tag{2.15}$$

where $\overline{\eta}$ is the detection efficiency of the reaction product averaged according to the chosen energy interval for its detection. Assuming a uniform distribution of hydrogen in a thin layer Δx and that the resonance shape has the Breit-Wigner form, the integration of (2.15) results in the desired expression for the concentration [16]

$$C(x_R) = N(E_0)S_1 \left[1.1\,N_0\Delta\Omega\frac{d\sigma}{d\Omega}(E_R)\overline{\eta}\Gamma n_{0M}\right]^{-1} \ . \tag{2.16}$$

Some basic ideas and the theory of depth profiling with narrow resonances are considered in detail by, e.g., *Amsel* and *Maurel* [17–20].

3. Analytical Characteristics of the Method

3.1 Detection Limit

The detection limit C_{min} is, according to [21], a minimum hydrogen content that may be detected with reliability $P(\alpha)$. When both the signal effect and the background have a Poisson distribution, as is usually true for nuclear-physics analysis, the detection limit is found according to the expression

$$C_{min} = \frac{K\alpha^2}{2}\left[1 + \left(1 + 8N_b/\alpha^2\right)^{1/2}\right] , \qquad (3.1)$$

where N_b is the background signal; K is a calibration factor that characterizes the counter response to hydrogen present in a sample when no dispersion of K occurs; α is the reliability coefficient ($\alpha = 1$ at $P = 68\%$ and $\alpha = 2$ at $P = 95\%$). If $C(x)$, the concentration distribution, is measured by the energy analysis method, then K is the term in square brackets in (2.3).

The background level N_b corresponds to the number of counts in the relevant energy spectrum region with hydrogen-free bulk material. The background yield, related to the reaction on a sample matrix, is determined in a similar manner to the reaction yield from hydrogen, i.e. by (2.1), where $n_0(x)$ and $d\sigma/d\Omega$ refer to the sample material. That is why the background level, and thus the hydrogen detection limit, depend on the relationship between the reaction cross-sections for hydrogen and those for the sample matrix (to be more exact, they depend on the shape of the relevant excitation functions) and may change with the analysis depth. There may exist some other background contributions, for example those from sources not connected with the beam striking the sample.

The resonance method also shows the contribution of hydrogen-beam interactions to the background at a depth $x \neq x_R$, i.e. due to a non-zero value for the off-resonance yield. A large difference between the resonance and off-resonance heights of excitation function implies that this reaction can give a signal only from hydrogen located at depth x_R. However, in contrast to (2.15), the calculation of the background is performed by integration over the whole energy range, which corresponds to the range of an incident particle in a sample material. Thus, the influence of the off-resonance region of an excitation function together with the "tails" of other resonances may be rather considerable. The sensitivity of the resonance method is further reduced by the ever-present surface hydrogen contamination as the latter makes the off-resonance yield increase. When using γ ray reaction spectrometry, the background radiation from natural and cosmic sources limits the minimum detectable hydrogen concentration.

Suppression of the various background contributions is ensured (1) by the choice of the energy and type of bombarding particles; (2) by use of a thin absorber or an electrostatic filter in front of the detector; (3) by space-time correlation; (4) by application of low-level counting techniques, etc. A nuclear reaction with a high positive Q value allows a considerable decrease of the background level by amplitude discrimination of detector signals.

According to the type of nuclear reaction used, as well as the sample material and the analysis conditions, the detection limit varies from 1 to 10^5 at.ppm. Quite often the term "detection limit" is replaced by "sensitivity", although the latter means the ability to detect a slight difference in concentration with a given reliability. It has been suggested that "detection limit" be used as a quantitative characteristic of the analytical method [22]. However, the second term is also used sometimes.

3.2 Rapidity

This aspect of the analysis has no clear definition; it shows how rapid an analysis is and is characterized by the time required to accumulate information on the reaction yield. According to (2.1) the analysis is more rapid in the cases of (1) large cross section; (2) high hydrogen content; (3) large number of particles striking a sample; (4) high detector efficiency and (5) large detector solid angle.

Sometimes the incident radiation may be limited by the source intensity, but more often it can not be increased because a high beam current may lead to a significant distortion of the initial hydrogen concentration distribution due to radiation defect formation and sample heating. A high beam current and large detector solid angle may result in the overload of the detection electronics. Moreover, the depth resolution may deteriorate for a large solid angle.

The rapidity of the analysis is important not only from the point of view of analytical laboratory efficiency. Long exposure times may result in lower accuracy and in resolution degradation due to non-stability of the radiation source and of the detection electronics. In the study of diffusion profiles, high rapidity of analysis makes it possible to widen the range of the diffusion coefficients measured.

3.3 Selectivity

This characteristic is related to the ability to obtain separately information on the content of various hydrogen isotopes. A strong isolated resonance in a cross-section, a high Q value or some other peculiarity of a nuclear reaction makes it possible to perform the analysis for hydrogen. But this condition is usually true only for a single isotope analysed. In the case of elastic scattering, which is characterized by a slight difference in cross-sections for the different hydrogen isotopes, the possibility of their simultaneous determination depends on the ability of a detecting system to identify the type of recoil nucleus and also the background level relevant to each particle type.

3.4 Accuracy

The accuracy of analysis addresses the quality of measurements and reflects the correspondence between the measured value and the actual value. High accuracy requires small values of random as well as of systematic errors. Random errors consist primarily of an error in the determination of the number of incoming and emitted particles, connected with the statistical nature of nuclear interactions and particle detection.

As a rule, nuclear-physics analysis provides the evaluation of the total number of events N and the number of background counts N_b to meet the condition $\Delta N = N - N_b \gg 1$. This implies the normal law of probability distribution for N. Hence, dispersion of N is calculated to $s_N = N^{1/2}$. To detect a difference between N and N_b with probability $P(\alpha)$ it is enough to meet the following conditions:

$$\Delta N = \alpha s_{\Delta N} = \alpha\sqrt{s_N^2 + s_{N_b}^2} = \alpha\sqrt{\Delta N + 2N_b} \quad \text{or} \tag{3.2}$$

$$\Delta N = \frac{\alpha^2}{2}\left[1 + \left(1 + 8N_b/\alpha^2\right)^{1/2}\right] . \tag{3.3}$$

According to (3.2) the relative error of "pure" signal ΔN, i.e.

$$\varepsilon = \sqrt{\Delta N + 2N_b}/\Delta N ,$$

is connected to $P(\alpha)$ by

$$\varepsilon\alpha = 1 . \tag{3.4}$$

This allows one to find the permissible error $\varepsilon = \alpha^{-1}$ for the registration of the "pure" signal with a given probability $P(\alpha)$ and, vice versa, for a given statistical error ε it is possible to obtain the probability that the signal will exceed the background, corresponding to the coefficient $\alpha = 1/\varepsilon$. Random errors also include those appearing as a result of variation in the geometry and instabilities in detection conditions on comparing samples and standards. In addition, signal loss due to the dead time of the electronics is also considered as a random error.

Systematic errors in hydrogen analysis by nuclear physics may result from: errors in determination of nuclear reaction cross section or standard hydrogen concentration; error in detector efficiency evaluation; the poor separation of the hydrogen-reaction effect from the background of competing processes; the influence of surface sample contamination; error in geometrical factor calculation; errors in stopping powers and ranges, etc.

The total error of the analysis may be found [21] from

$$\varepsilon = \varepsilon_s + \alpha\varepsilon_r = \sum_i \varepsilon_{si} + \alpha\sqrt{\sum_i \varepsilon_{ri}^2} , \tag{3.5}$$

where ε_s and ε_r are systematic and random errors, respectively; α is a reliability coefficient, determined by a given confidence level. When the yield from a sample is compared to that from a hydrogen standard [see (2.4)], the hydrogen content thus found, avoids some of the errors mentioned. The main contribution in this case is from statistical fluctuations of counting rate in the measurements of the reaction yield, the background, and the incident beam onto both the sample and the standard, and also from errors in the tables of ranges and stopping powers.

The accuracy in hydrogen content in the standard should also be taken into account.

As the errors in n_{Hs} and dE/dx are usually comparatively small, the total relative error may be calculated in a more simple way, i.e. as the square root of the sum of squares of relative errors ε_i for each value in (2.4),

$$\varepsilon_c = \alpha \sqrt{\sum_i \varepsilon_i^2} \ . \tag{3.6}$$

For the determination of the concentration distribution $C(x)$ it is important to obtain correct spectrometric information and to choose the correct method of energy-depth transformation. The corrections for counting losses due to multiple scattering should be taken into account. Uncertainties in the material density may also contribute to uncertainties in the depth. And, finally, the accuracy of concentration profiling depends on our knowledge of the depth resolution of the experimental technique used.

3.5 Depth Resolution

The nuclear physics methods that allow evaluation of not only the total hydrogen content but also its distribution in a material are characterized by the depth resolution attainable. We will consider one-dimensional distributions since most nuclear physics techniques have been very successful in determining the concentration vs depth, but have not been able to measure the lateral distribution of hydrogen.

The coordinate x that indicates the location of hydrogen in a material is evaluated with uncertainty Δx that depends on the nature of the nuclear reaction, sample material properties, and the detection parameters. The latter are the energy of the probing particles, their angle of incidence, the angle of observation, the detector energy resolution, etc. In order to allow a correct interpretation of the depth profiles extracted by means of the nuclear reactions, a detailed analysis of the depth resolution is indispensable in every case.

The relationship between the measured concentration distribution $C(x)$ and the real one $C_r(x)$ is determined by

$$C(x) = \int\limits_0^l C_r(x') f(x, x') \, dx' \tag{3.7}$$

where $f(x, x')$ is the resolution function which may be found by appropriate calculations or, if possible, by direct experimental measurements. The full width at half maximum (fwhm) of this function Δx characterizes the depth resolution.

Let us consider now a more complicated case where the concentration profile is obtained using a reaction with charged particles in the entrance and exit channels. Differentiating (2.6) or (2.7) we get the stopping parameter

$$\frac{dE}{dx} = - \left[\frac{\partial E_3}{\partial E_1} \frac{S_1}{\sin \psi} + \frac{S_2}{\sin(\theta - \psi)} \right] = -\tilde{S} \; , \tag{3.8}$$

which is necessary for the profile calculation of (2.3). From this it follows that the depth resolution

$$\Delta x = -\Delta E \, \tilde{S}^{-1} \; , \tag{3.8'}$$

depends on the stopping power of the hydrogen-containing material and on the geometry, as well as on the nuclear reaction kinematics (factor $\partial E_3 / \partial E_1$). Other conditions being the same, the resolution is intrinsically better in reflection geometry than in transmission geometry, as in the latter case the second term in (3.8) is negative. It follows from (3.8) that, for a given energy resolution ΔE, the depth resolution Δx changes with depth according to the energy dependence of the stopping power. However, ΔE is also depth-dependent, which is why the overall resolution should be studied in each particular case.

Particles originating from depth x, recorded by a detector with mean energy $E(x)$, show energy fluctuations due to statistical effects characterizing the incoming particles, the detection and the slowing down process, i.e. energy straggling and multiple scattering of particles as they traverse the material. All these contributions, once determined, must be combined for the evaluation of the total resolution.

Now let us consider the factors affecting the energy resolution:

1) ΔE_0: the energy spread of the incident particles hitting the target;

2) ΔE_d: the detector resolution coupled with the electronics' noise contribution;

3) ΔE_g: the geometric spread, i.e. the effect of the finite angular aperture of the experimental set-up, inducing an uncertainty in the detection angle $\Delta \theta$ that leads to the relevant kinematic expansion $\Delta E_g = (\partial E / \partial \theta) \, \Delta \theta$. Using (2.6) we get

$$\Delta E_g = \left[\frac{\partial E_3}{\partial \theta} - S_2 x \frac{\cos(\theta - \psi)}{\sin(\theta - \psi)} \right] \Delta \theta \; , \tag{3.9}$$

where

$$\partial E_3 / \partial \theta = \pm 2 \, E_3 P^{-1/2} \sin \theta \tag{3.10}$$

is obtained as a derivative of (2.8). Geometric spread may sometimes provide a considerable contribution to energy resolution. Thus, hydrogen profiling by the nuclear recoil method (elastic scattering) yields $\partial E_4 / \partial \theta$ of 350 keV/degree for $\theta = 45°$ and $E_4 = 10$ MeV, whereas a surface-barrier detector resolution of several tens of keV may be easily achieved;

4) ΔE_{st}: the effect of energy straggling due to statistical fluctuations of energy losses of particles, both incoming and outgoing, as they penetrate the sample material from the surface to the reaction point and then from this point to the exit. The contribution of straggling may be calculated as follows [23]:

$$s_{st} = a^2 L(\chi)/2 \quad \text{for} \quad \chi \leq 3$$
$$= a^2 \qquad \text{for} \quad \chi \geq 3 \ , \tag{3.11}$$

where

$$a^2 = 4\pi z^2 Z e^4 n_0 x \ ; \quad L(\chi) = S(m_e/4z^2\hbar^2 n_0)\chi \ ; \quad \chi = v^2/(e^2/\hbar)^2 Z \ .$$

$\Delta E_{st} = 2.36\, s_{st}$; s is the standard deviation; m_e is the electron mass; z and v are particle charge and velocity, respectively; n_0 and Z are the atomic density and the charge of the sample material, respectively;

5) ΔE_m: the energy spread due to angular straggling as a result of multiple scattering. Travelling through the material, incident and emitted particles can change their direction due to the great number of Coulomb scattering events. This effect is characterized by the mean-square angle of multiple scattering $\overline{\vartheta}^2$, which depends on the particle type (z, m) and velocity v, on the properties of the stopping medium (Z, A, n_0) and on the penetration depth x [24]:

$$\overline{\vartheta}^2 = K \ln \left[4\pi Z^{4/3} z^2 n_0 x \, (\hbar/mv)^2 \right] \ , \tag{3.12}$$

where

$$K = 0.157\, Z\,(Z+1)z^2 x/(pv)^2 A \ .$$

As a result of multiple scattering, the angular distribution of particles defines an additional spectrum of observation angles of width $\Delta\theta$. Hence, the angular spread acts through the kinematics like the geometrical spread. Thus, in accordance with (3.9), the outcoming particles have an additional energy spread ΔE_m. In addition, multiple scattering leads on the one hand to the certainty that some particles initially scattered into the detector solid angle will not reach the detector. On the other hand, particles emerging from a collision at an angle falling outside the limits of the detector aperture may be detected. This effect also results in deterioration of the energy resolution. Multiple scattering causes the most important energy expansion effect at large sample thickness (more precisely, at large energy losses). It has been suggested in [25] that experimental data on this spread be used to estimate multiple scattering of charged particles;

6) ΔE_l: the energy spread relevant to the lateral spread $\Delta x_\perp$ of the particle trajectories (incident and emerging) due to particle deflection as a result of multiple scattering. Figure 3.1 [26] shows the multiple scattering effect for an incident beam. It is seen that the particles crossing a plane at depth x acquire not only angular spread but also uncertainty in path lengths,

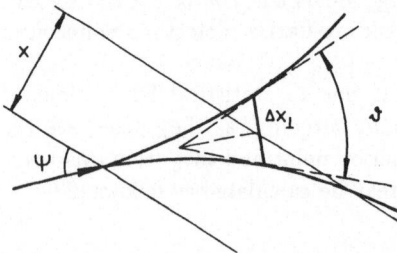

Fig. 3.1.
Multiple scattering effect on probing ions

$$\Delta l_1 = \Delta x_\perp \tan\psi \ . \tag{3.13}$$

For outgoing particles this spread may be written as

$$\Delta l_2 = \Delta x_\perp \tan(\psi - \theta) \ . \tag{3.14}$$

The effect is expressed by the corresponding energy expansion

$$\Delta E_1 = S\Delta l \ . \tag{3.15}$$

As multiple scattering causes considerable deterioration of depth resolution, theoretical techniques [27] for estimating the lateral spread $\Delta x_\perp$ (and for its relation with angular spread) have been developed, as well as methods for its experimental determination [28]. It should be noted that the geometric spread mentioned above, as well as sample surface roughness, also lead to a spread in path lengths.

7) ΔE_a: an additional energy broadening arises from straggling in an absorber foil if used in front of a detector to eliminate the high background of short-range projectiles. It is reasonable to employ such a filter if a heavy ion beam is used, since its elastic scattering on a sample matrix leads to electronic overload and to the rapid damage of a detector.

These contributions all depend on x, [except that 1, 2 and 7] and, strictly speaking, are not independent of each other. Nevertheless, a simplified procedure has been used in many studies: the total energy resolution ΔE_t is calculated assuming that individual processes which influence the resolution may be summed quadratically. That is true only when the contributions are fully independent and they follow the Gaussian distribution. In other words, $\Delta E_i = 2.36\, s_i$, the resultant standard deviation $s_t^2 = \sum s_i^2$ and the energy resolution function is

$$F(E, E') = \exp\left[-(E - E')^2/2s_t^2\right]\Big/\sqrt{2\pi s_t^2} \tag{3.16}$$

with fwhm $\Delta E_t = 2.36 s_t$. The instrument energy spectrum $N(E)$ is connected to the initial one $N'(E')$ by the convolution

$$N(E) = \int N'(E')F(E, E')\, dE \ . \tag{3.17}$$

Then the resolution function of interest may be written as

$$f(x, x') = F(E, E')\, dE/dx \ , \tag{3.18}$$

where dE/dx is determined by (3.8).

This subject is discussed in detail in [29], where a more complete method for the overall resolution calculation is suggested. Figure 3.2 (taken from [29]) shows (for example) depth dependencies obtained for various contributions as well as for the total depth resolution of depth concentration profiling of deuterium using the $D(^3He, p)\,^4He$ reaction by detecting backscattered protons.

In order to obtain the real hydrogen depth-distribution it is necessary to solve the integral equation (3.7) on the basis of knowledge of $f(x, x')$ and the measured

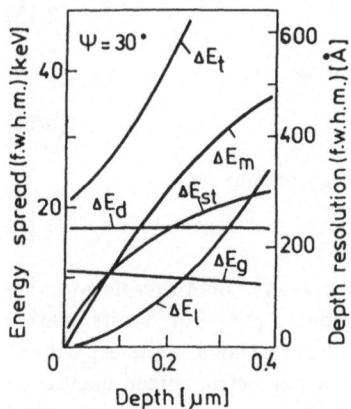

Fig. 3.2. Dependence of energy spread and depth resolution on depth in amorphous silicon, calculated for deuterium profiling using the D(^{3}He, p) ^{4}He reaction. $\theta = 165°$; $E_{3_{He}} = 700\,\text{keV}$; $\psi = 30°$

concentration profile. The solution of this deconvolution problem is discussed, for example, in [30, 31]. However, sometimes the aim of an investigation is the determination of physical values that appear as parameters in a theoretical model for the calculation of concentration profiles in certain processes. In this case a simple convolution (3.7) of the calculated concentration profiles is sufficient and the parameters of interest may be obtained by comparing this result with the experimental distribution, i.e. by a fitting procedure.

The energy spectrum of a reaction product does not seem to be of any interest in nuclear resonance reaction techniques as a hydrogen depth-profile in a sample is evaluated only by using the energy and the stopping power of an incident particle [see (2.14)]. Instead of (3.8) we have, from (2.14),

$$\Delta x = \Delta E_t \sin \psi / S_1 \quad , \tag{3.19}$$

where

$$\Delta E_t = \left[\Gamma^2 + (\Delta E_0)^2 + (\Delta E_s)^2 \right]^{1/2} \quad . \tag{3.20}$$

Γ is the resonance width; ΔE_0 is the spread in the incident beam energy including the Doppler broadening; ΔE_s is the spread due to stopping processes of the beam particles in the sample.

In shallow concentration profiling ($\Delta E_s = 0$) only the resonance width influences the resolution since the Van-de-Graaff or Tandem accelerators used for resonance techniques are characterized by a low energy spread of ion beams (few keV). However, if the resonance is very narrow (as in the case in the ^{1}H(^{15}N, $\alpha\gamma$) ^{12}C reaction) the resolution is limited by the Doppler broadening [32] due to thermal motion of hydrogen atoms in solids.

The effective energy of a bombarding ion E_0' seen by the vibrating hydrogen atom at the start of the reaction will be

$$E_0' = m_1(v_0 - v)^2/2 = E_0 - m_1 v_0 v \cos \alpha + m_1 v^2/2 \quad , \tag{3.21}$$

where m_1 is the incident ion mass and v is the thermal motion rate vector that makes an angle α with the velocity vector of an incident particle v_0. The last term in (3.21) has the value of the order of kT and is, hence, quite negligible. The second term is random. This leads to the Doppler broadening that follows the Gaussian law of variance with dispersion:

$$s_D^2 = 2\,m_1 E_0 kT/m_2 \ , \tag{3.22}$$

where m_2 is the hydrogen mass. The evaluation in [32] for the reaction ^{1}H(^{15}N, $\alpha\gamma$)^{12}C with ^{15}N ion energy $E_0 = 6.5\,\text{MeV}$ showed broadening due to the Doppler effect of $\sim$5 keV fwhm at 300 K which exceeds the resonance width ($\Gamma = 1.8\,\text{keV}$). Hence, further resolution improvement requires sample cooling, which, in its turn, implies the use of high quality, low vacuum target chambers.

Depth resolution for different methods and sample materials varies from tens of microns to tens of Angströms, although the highest resolution of $\sim$40 Å was obtained, using ^{15}N ions, only for surface hydrogen in Si.

3.6 Analysable Depth

The maximum thickness, $l_{\max}$, of a sample for which the bulk profile may be measured depends on (1) the properties of the reaction chosen, its Q value and the excitation function type; (2) the energy of the probing particles; (3) the stopping power of the sample material; (4) detection conditions, i.e. geometry, background, angular resolution, energy threshold of a spectrometer, etc. For example, in the case of hydrogen depth-profiling by elastic-recoil-detection, the maximum analysable depth is calculated using (2.13), where $R_p(E)$ and θ are replaced by $R_p(E_{th})$ and by $\theta + \Delta\theta$, respectively (E_{th} and $\Delta\theta$ are the energy threshold of a proton spectrometer and the detector angular aperture, respectively).

In analysis by nuclear-resonance reaction, the profiling depth is particularly limited by complications due to the superimposed contributions of other resonances. Depending on the reaction type and the experimental conditions, the profiling depth may be varied from several micrometers to several millimeters.

3.7 Choice of Nuclear Reaction

The various nuclear physics methods for the determination of hydrogen and its isotopes are listed in Table 3.1, which summarizes the data published since 1967. Nuclear techniques for hydrogen determination can be divided into five groups. The first group includes reactions used to evaluate total hydrogen content, while the remaining four represent methods to obtain concentration-depth distributions. The symbols T, R and E designate the methods of total content, resonance and energy analysis, respectively.

It seems reasonable to divide the energy analysis methods into three groups, as there are distinct differences, both in the principle and in the method, in hydrogen analysis using nuclear reactions, heavy ion elastic scattering and nucleon elastic scattering. It should be noted that the group of nuclear recoil methods may be expanded, as there are no practical limitations to the incident ion type apart from the possibility of its acceleration.

Table 3.1. Summary of nuclear physics methods for hydrogen determination

Nuclear Interaction	Detected particle	Technique (E_R [MeV])	Detection limit (material) [at. H/at. M]	Depth resol./analys. depth (material) [μm]	References
1	2	3	4	5	6
$^2H(\gamma,n)^1H$	n	T	2×10^{-6} (H_2O)		[33]
$^2H(^{15}N, p\gamma)^{16}O$, $^2H(^{15}N, n\gamma)^{16}O$	γ	T	5×10^{-5} (Si)		[34]
$^1H(n, n)^1H$	n	T	10^{-3} (Zr)		[38]
$^1H(n, \gamma)^2H$	γ	T	10^{-3} (Zr)		[38]
$^1H(n, p)n$	p	T	10^{-3} (Si)		[35]
$^1H(n, n)^1H$ + activation	γ	T	10^{-4} (Ba, Mo)		[42]
$^2H(n, d)n$ + activation	γ	T	10^{-3} (H_2O)		[43]
$^1H(n, p)n$ + x-ray radiation	X	T	10^{-1}		[44]
$^1H(^{15}N, \alpha\gamma)^{12}C$	γ, α	R (13.35)	10^{-3} (Si)	$1 \times 10^{-2}/1$ (Si)	[45]
		(6.4)	10^{-5} (Si)	$4 \times 10^{-3}/4$ (Si)	[19, 51, 52]
$^1H(^{19}F, \alpha\gamma)^{16}O$	γ	R (16.44)	2×10^{-3} (Si)	$3 \times 10^{-2}/0.4$ (Si)	[45, 56]
		(6.42)	10^{-2} (Si)	$1.5 \times 10^{-2}/1.2$ (Si)	[45, 56]
$^1H(^7Li, \gamma)^8Be$	γ	R (3.07)	10^{-3} (Si)	$1 \times 10^{-1}/8$ (Si)	[45, 58]
$^1H(^{11}B, \alpha)^8Be$	α	R (1.79)	10^{-2} (Si)	$8 \times 10^{-2}/0.1$ (Si)	[9, 45]
$^1H(^{18}O, \alpha)^{15}N$	α	R (11.25)	10^{-4} (Si)	$2 \times 10^{-2}/3.5$ (Si)	[45]
$^1H(^{13}C, \gamma)^{14}N$	γ	R (22.55)			[45]

1	2	3	4	5	6
$^2H(^3He, p)^4He$	p, α	E	10^{-3} (Si)	$2 \times 10^{-2}/0.3$ (Si)	[8, 29]
$^2H(d, p)^3H$	p	E	10^{-2} (Ni)	0.07/4 (Ni)	[26]
$^1H(t, n)^3He$	n	E	2×10^{-1} (Ti)	0.5/50 (Ti)	[78]
$^2H(d, n)^3He$	n	E	10^{-3}	1/45 (Ti)	[78]
$^3H(p, n)^3He$	n	E	10^{-3}	1/45 (Ti)	[78]
$^3H(d, α)n$	α	E	1.4×10^{-4} (Ti)	$4 \times 10^{-2}/1.4$ (Ti)	[67, 81]
$^1H(^3He, p)^3He$	p	E	10^{-3} (Si)	$8 \times 10^{-3}/0.5$ (Si)	[92]
$^1H(^4He, p)^4He$	p	E	10^{-3} (Al)	$5 \times 10^{-2}/3$ (Al)	[84, 90, 95]
$^1H(^{14}N, p)^{14}N$	p	E	10^{-4} (Nb)	$10^{-2}/0.2$ (Nb)	[85]
$^1H(^{32}S, p)^{32}S$	p	E	2×10^{-4} (Si)		[86]
$^1H(^{35}Cl, p)^{35}Cl$	p	E			[88]
$^1H(p, 2p)$	2p	E1	10^{-6}*	14/170 (Ti)	[97, 99]
$^1H(n, p)n$	p	E	0.1*	37/700 (Ti)	[101]
$^2H(n, d)n$	d	E1	10^{-3}*	23/290 (Ti)	[101]
$^3H(n, t)n$	t	E	10^{-3}*	6.5/110 (Ti)	[101]

* There is almost no dependence on the material studied

The reaction presentation shows which hydrogen isotope is determined (the first symbol), which incident particles are used for the analysis (the first symbol in brackets) and which particle is detected (the second symbol in brackets). Since one or the other, or both, particles may be registered from the same reaction as well as the secondary reaction products, this information is placed in a separate column. The table also presents the main characteristics of the methods: detection limit, depth resolution, maximum probing depth. The last two values are given as a ratio in the same column, because their relationship determines the relative depth resolution of the method. The symbol in brackets corresponds to the hydrogen-containing material. If there is no symbol in brackets it means either the independence of the given characteristic on material type, or the absence of this data in the corresponding publication.

It should be noted that the cited values are the best obtainable for the given method (the resolution is given for the near-surface region) and they are approximate because these values may differ under specific experimental conditions. Moreover, in some publications, uncertainties in the description do not permit assessment of the accuracy. In the text the detection limit is given in the units used by the authors, but is recalculated in the summary table thus providing a more convenient unit, i.e. at.H/at.M. The summary of nuclear reaction techniques shows that if a suitable source of radiation is available, the investigators have many ways to detect hydrogen directly in studies of its behaviour.

Increasing the incident particle mass and decreasing its energy, and choosing a heavier product for registration (i.e. increasing the stopping power), lead to the improvement of the absolute depth resolution but also to the decrease of the analysable depth. Methods using monochromatic neutrons or protons accelerated to large energies can be used to obtain larger analysable depths.

The improvement in depth resolution is limited by the resonance width in resonance reactions, whereas in energy analysis it is determined by the energy resolution of the spectrometer used. However, in deep-region analysis the resolution is limited in both methods by the statistical processes accompanying the passage of charged particles through material. In the investigation of the near-surface region, the resonance reaction $^1H(^{15}N, \alpha\gamma)\,^{12}C$ seems to be most suitable for high resolution.

To obtain the smallest detection limit it is best to chose a reaction for which the relative cross sections of the competing processes is small, for example $^2H(^3He, p)$, or to suppress the background by shielding $[^1H(^{15}N, \alpha\gamma)\,^{12}C]$ or by performing the correlation measurements $[^1H(p, 2p)]$. On the contrary, when studying materials with high hydrogen content it is convenient to use heavy-ion recoil methods or a simple method of monochromatic neutron elastic scattering.

The choice of method should not only provide the analysis characteristics of interest, but also take into account any possible influence of a bombarding beam on the hydrogen concentration distribution studied. In this respect energy analysis methods appear to be preferable, as they allow overall concentration profiling with one set of measurements. It is also desirable to use light particles and to exclude the region at the end of a particle range, as this region is characterized by a high degree of radiation damage.

4. Determination of Total Hydrogen Content

In cases where the incident and emerging particles have no charge, or if the particle energies are not measured, only the total hydrogen content can be determined. These reactions are generally considered to be methods for the determination of the total hydrogen content, in spite of the fact that they appear to be particularly powerful for local concentration or spatial distribution measurements by a sequential analysis using a narrow particle-beam or position-sensitive detector.

If the total hydrogen content is evaluated by neutrons or γ rays, then the cross-section $d\sigma/d\Omega$ is constant throughout a whole sample. Integrating (2.1) from $x = 0$ up to $x = l$, we obtain the expression for the total reaction yield

$$N = N_0 \overline{C} \overline{\eta} \frac{d\sigma}{d\Omega}(E_0, \theta) l \Delta\Omega \, n_{0M} \; . \tag{4.1}$$

Here $\overline{\eta}$ and $\Delta\Omega$ are the mean detection efficiency and detector solid-angle, respectively. This expression can be used to determine the average concentration of hydrogen $\overline{C}$ in a sample of small thickness l, because the decrease in the intensity between incident and emerging radiation, due to the interaction with the sample material itself, is not taken into account.

4.1 The ^{2}H$(\gamma, n)\,^1$H Reaction

This reaction is characterized by the relatively low energy threshold (2.23 MeV). As only the ^{9}Be nucleus is known to have a lower threshold [1.67 MeV for the (γ, n) reaction], the deuterium content may be determined by neutron detection in almost any substrate. Various radionuclides, as well as bremsstrahlung from electron accelerators, may be used as sources of γ radiation. A γ ray energy of 4.4 MeV corresponds to the maximum reaction cross-section (2.5 mb), which is why it is reasonable to use an electron accelerator of $\sim$5 MeV.

The method has been applied mainly for the analysis of deuterium in water with the very low detection limit of about 2×10^{-4} at.% (natural content is 1.5×10^{-2} at.%) [33]. It may be used for monitoring technological processes of hydrogen isotopes separation. The method can also be applied to studies where deuterium atoms are used as labels, and in investigations of geochemical anomalies of deuterium content. The reaction may also be used to evaluate moisture content in a material by natural deuterium impurity detection.

4.2 The ^{2}H$(^{15}$N, p$\gamma)\,^{16}$N and ^{2}H$(^{15}$N, n$\gamma)\,^{16}$O Reactions

^{15}N ions are widely used for hydrogen detection by means of the ^{1}H$(^{15}$N, $\alpha\gamma)\,^{12}$C reaction. This is a well known technique specific to hydrogen resonance profiling; therefore it will be discussed in Sect. 5.1.

However, recently *Hayashi* et al. [34] have used the ^{15}N-induced nuclear reactions as a new method for the high sensitivity analysis of deuterium in solids. These reactions are as follows: ^{2}H$(^{15}$N, p$\gamma)\,^{16}$N and ^{2}H$(^{15}$N, n$\gamma)\,^{16}$O. The ^{16}N nucleus produced by the first reaction decays to the excited states, as well as to the ground state of the ^{16}O by β decay. Therefore photopeaks at 2.75, 6.13 and 7.12 MeV, together with the corresponding single escape (s.e.) and double escape (d.e.) peaks in Fig. 4.1, are assigned to originate from the excited states of ^{16}O produced by both the first and the second reactions. Since the excitation functions of these reactions are quite similar in shape and have no resonance below 16 MeV, they cannot be applied to resonance depth profiling, but nevertheless are useful for the determination of the total amount of deuterium by measuring the intensity of the γ rays in the range 5–8 MeV.

As is understood from Fig. 4.1, two hydrogen isotopes can be measured simultaneously without mutual interference. However for ^{15}N ion energies greater than 14.45 MeV, the interference reaction channel ^{2}H(^{15}N, $\alpha n \gamma$) ^{12}C is opened, giving from deuterium γ rays with the same energy as is emitted from hydrogen.

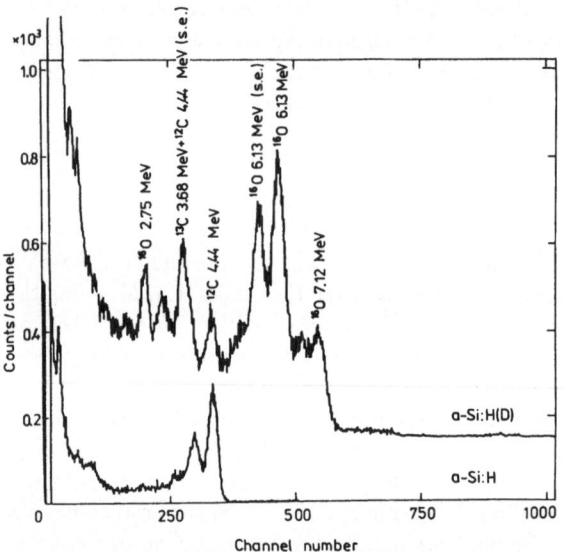

Fig. 4.1. γ ray spectra recorded from a-Si:H(D) (*upper curve*) and from a-Si:H (*lower curve*), induced by 7 MeV ^{15}N ions

As the background in the high-energy part of the γ ray spectrum is rather low, deuterium content can be measured down to 10^{18} at.cm^{-3} (or 5×10^{-5} at.D/at.Si), which is quite low. However, high numbers of neutrons with continuous energy are emitted from the breakup process (^{15}N→^{14}N+n) over about 11 MeV. This demands the use of pulse-shape discrimination techniques for n–γ separation. Also, above 11.9 MeV, γ rays of the same energy are released from the excited-state of ^{16}O excited by the inelastic scattering of ^{15}N ions by oxygen absorbed on the surface of most materials. In spite of the fact that the cross-section of inelastic scattering was determined to be 10^3–10^{-4} times smaller than that of deuterium-originated reactions, this process may sometimes limit the sensitivity.

The large cross-section of the $^9\text{Be}(^{15}\text{N},^{16}\text{N})\,^8\text{Be}$ reaction is not so important for deuterium analysis, as beryllium is rarely present in most materials.

Taking into account everything said above, the authors [34] conclude that the most suitable energy for ^{15}N ions for the general purpose of deuterium analysis using a simple measurement system is about 10 MeV.

4.3 Neutron Methods for Hydrogen Determination

In the past [35] hydrogen in ore was determined by the scattering of neutrons from an isotopic source or a neutron generator. A sample was placed close to a detector and recoil protons were detected. Since the proton energy E_p depends very much on a scattering angle θ, as $E_p = E_n \cos^2 \theta$, even in the case of monochromatic neutrons it varies from a maximum equal to neutron energy E_n down to zero. That is why energy analysis of recoil protons is not useful, since only the total hydrogen content in a sample may be evaluated. The authors give the detection limit for quartz, magnetite and galenite $\sim 10^{-3}$ wt.%.

The use of pulsed neutron beams increased the efficiency of neutron methods of hydrogen determination [36]. *Overley* [37] used pulsed beam and time-of-flight techniques to measure the energy dependence of a collimated fast-neutron continuum transmitted through a hydrogen-containing sample. Time-of-flight spectra obtained by scanning of the sample allowed the determination of the spatial distribution of hydrogen and other elements in a plane of the object using the technique of computed tomography.

The high thermal-neutron cross-section of hydrogen gives rise to a strong attenuation of a neutron beam in transmission, compared with the absorption of the sample for most metals. The transmitted intensity I is given by an exponential law

$$I = I_0 \exp\left[-n_{0\text{M}} l (\sigma_\text{M} + C\sigma_\text{H})\right] \quad , \tag{4.2}$$

where I_0 is the intensity of an incident neutron beam; σ_n and σ_H is the cross-section of a sample matrix and hydrogen respectively and C is the hydrogen concentration expressed in at.H/at.M. Thus, the attenuation I_0/I of a transmitted neutron beam may be a convenient tool to determine the hydrogen concentration in the bulk of a sample. In a series of experiments (see, for example, [38–41]), either the scanning of a sample with a narrow thermal-neutron beam, or neutron radiography was used to obtain the lateral hydrogen distribution.

The scattering $^1\text{H}(n,n)\,^1\text{H}$ and the capture $n + {}^1\text{H} \rightarrow \text{D} + \gamma$ of reactor-neutrons were used to determine hydrogen in zircaloy and yttrium samples [38]. The experimental device is shown schematically in Fig. 4.2.

In order to minimize the error in the determination of hydrogen concentration, the main energy region of the neutrons is limited by filtering the neutron beam in an appropriate way. The use of epithermal neutrons leads to an isotropic angular distribution of the neutrons scattered from metal atoms and to an anisotropic distribution when a scattering event occurs on hydrogen atoms. Therefore, neutron detectors (BF$_3$ counters or fission chambers) placed on both sides of the sample distinguish between neutrons scattered by the sam-

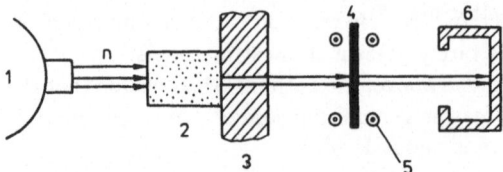

Fig. 4.2. Neutron scattering device for the hydrogen determination; (*1*) reactor neutron beam, (*2*) neutron transmitter, (*3*) collimator, (*4*) sample, (*5*) neutron detectors, (*6*) beam catcher

ple matrix and those by hydrogen. Such detector location also eliminates the influence of varying composition and thickness of samples. The absolute content was obtained by measuring count-rates of several samples with different known-hydrogen-content.

Since the thermal capture cross-section is known with high accuracy, an absolute direct determination of hydrogen concentration can be performed using another method based on the measurement of the 2.2 MeV capture γ rays. Both methods give results that are in good agreement within the experimental error. The authors [38] do not give numerical values of the detection limit for hydrogen determination, but most probably it is of the order of tenths of an atomic per cent.

Neutron radiography entails placing a sample and a converter/film cassette in a neutron beam. The latter includes a metal foil which emits x rays upon the absorption of neutrons, and a sheet of x ray film which is in the close contact with it. The blackening pattern of a developed x ray film is the neutron radiograph, which may be analysed on a microdensitometer. A densitometer scan is a direct measure of the variation of hydrogen concentration in a plane of the sample.

Hawkesworth et al. [39] have applied neutron radiography to reveal the distribution of hydrogen in electrolytically-hydrogenated palladium. It was shown that there is a tendency for the β phase ($C \sim 0.7$ at.H/at.Pd) to contract rather than to break up randomly. As for the sensitivity of this method, it was found that the α phase ($C \sim 0.02$ at.H/at.Pd) could not be detected in specimens which were thinner than 1 mm.

Neutron radiography was also used in the studies of the boundary that separates hydrogenated and non-hydrogenated regions in half-charged palladium [40], as well as of hydrogen-concentration fluctuations below and above (α–α') phase transitions in NbH$_x$ [41].

4.4 Determination of Hydrogen by Detecting Secondary Reaction Products

If the products of the primary interaction of nuclear radiation with hydrogen can not be registered, then their collision with nuclei of another element in the sample (indicator) can be used for the determination of hydrogen. The secondary reaction yield is:

$$N/N_0 \sim \sigma_H C_H \sigma_i C_i \ , \tag{4.3}$$

where σ_H and σ_i are the cross-sections of an interaction of primary radiation with hydrogen and an interaction of secondary radiation with an element-indicator, respectively; C are concentrations of hydrogen and an indicator in a sample.

Let us suppose that a sample contains, besides hydrogen, some other element with a large concentration of C_i and a large activation cross-section for thermal neutrons. Since hydrogen is considered to be a good neutron moderator, irradiation by fast neutrons leads to the result that the indicator element radioactivity will depend on the hydrogen content in a sample.

As an example, Fig. 4.3 [42] shows γ spectra of ^{139}Ba and ^{99}Mo obtained after irradiation of two pairs of samples in a flux of resonance and fast reactor neutrons: (a) barium peroxide and barium hydroxide; (b) metal molybdenum and ammonium molybdenum. Hydrogen-free and hydrogen-containing samples were chosen for having equivalent amounts of indicators, i.e. barium and molybdenum. It can be seen from Fig. 4.3 that the presence of hydrogen leads to an increase of intensity of induced radioactivity. The method has a sensitivity of up to 10^{-4} wt.%.

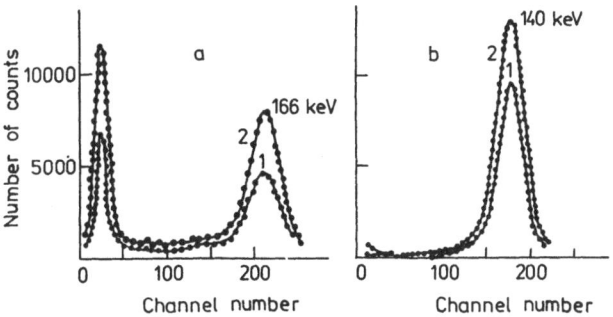

Fig. 4.3. γ ray spectra of ^{139}Ba (a) and ^{99}Mo (b) after neutron irradiation; curves 1 are related to barium peroxide (a) and metal molybdenum (b); curves 2 — to barium hydroxide (a) and ammonium molybdenum (b)

There are some references in [43] to the analysis of deuterium in water samples by detecting the radioactivity of ^{17}F nuclei ($T_{1/2} = 66$ s, $E_\gamma = 0.51$ MeV). Those nuclei appear in the ^{16}O(d, n) ^{17}F reaction due to recoil deuterons from sample irradiation by neutrons (from a reactor or a neutron generator). According to this method, the detection limit for deuterium is tenths of an atomic per cent.

The authors [44] suggested a method of hydrogen determination based on the fact that the recoil protons ionize the inner atom shells in the sample and thus cause characteristic x ray radiation. The detection limit of 0.5 wt.% was achieved due to the large cross-section of x ray radiation excitation (up to 10^3 b).

5. Hydrogen Concentration Depth Profiling

The capability to measure the hydrogen concentration distribution of a function of depth (depth profiling) by non-destructive methods appears to be one of the most important advantages of nuclear physics methods of analysis.

As mentioned in Chap. 2, depth profiling methods are divided into nuclear-resonance-reaction techniques and non-resonance ones. The latter are called the energy analysis methods and are based on both nuclear reaction proper ($Q \neq 0$) and on elastic recoil scattering ($Q = 0$).

5.1 Depth Profiling by Resonance Methods

Resonance methods for depth profiling, sometimes called NRRA (nuclear-resonance-reaction analysis), are based on nuclear reactions characterized by an excitation function with resonance structure, i.e. the energy dependence of a reaction cross-section is characterized by isolated, narrow and high peaks. The efficiency of a method depends on the characteristics of the reaction excitation function: (1) the width of the resonance used determines the depth resolution; (2) the relative areas under resonant and non-resonant parts of the excitation function define the sensitivity; and (3) the presence of adjacent resonances or increased non-resonant yield at energies $E_0 > E_R$ limits the maximum probing depth. The hydrogen concentration vs depth is obtained by measuring the reaction yield vs energy (2.15). This is performed step by step at an incident particle energy E_0 equal to or greater than the resonance energy at which the nuclear reaction can occur, i.e. by "shifting" the resonance through the sample according to (2.14). The choice of a nuclear reaction for resonance hydrogen depth-profiling was made relative to the inverse resonance reactions which were commonly used to determine the isotopes of light elements by proton bombardment:

$$^7\text{Li}(p,\gamma)\,^8\text{Be}, \qquad ^{15}\text{N}(p,\alpha\gamma)\,^{12}\text{C},$$
$$^{11}\text{B}(p,\alpha)\,^8\text{Be}, \qquad ^{18}\text{O}(p,\alpha)\,^{15}\text{N},$$
$$^{13}\text{C}(p,\gamma)\,^{14}\text{N}, \qquad ^{19}\text{F}(p,\alpha\gamma)\,^{16}\text{O}.$$

Thus, the reactions induced by the heavy-ion bombardment ($^7\text{Li}, ^{11}\text{B}$, etc.) of hydrogen allow the study of the hydrogen-concentration depth profile in the near-surface region of solids. The main characteristics of six resonant nuclear reactions have been recently presented and compared in [45], but the value for the first resonance width of the $^1\text{H}(^{15}\text{N}, \alpha\gamma)\,^{12}\text{C}$ reaction appears to be obsolete.

By analogy with the energy dependence of the cross-section $\sigma(E)$ that is called an excitation function, the reaction yield-dependence on incident particle energy $N(E_0)$ is sometimes called an excitation curve. In order to convert this excitation curve to a plot of absolute hydrogen concentration vs depth, it is necessary to use a relationship similar to (2.16);

$$C(x) = \text{const} \cdot N(E_0)S_1/N_0 \ , \tag{5.1}$$

where N_0 is the product of the ion-beam current and the exposure time; S_1 is the stopping power of the ^{15}N ions. There is no need to know the resonance cross-section, nor the efficiency of the detector if the constant of (5.1) is deduced by using a hydrogen standard, i.e. by normalizing the measured profile to that of a sample containing a known amount of hydrogen.

The hydrogen-concentration accuracy obtainable depends on the knowledge of, and the stability of, the standard hydrogen concentration during ion bombardment under ultra high vacuum (UHV) conditions. Besides, the standard material should be of a well-defined chemical composition and density, necessary for accurate calculation of S_1. It is desirable to have a constant hydrogen concentration over a large depth of the material.

A scan of the literature undertaken by *Whitlow* et al. [46] shows that minerals, hydrides, plastic and hydrogen implanted silicon or Al_2O_3 have been employed as hydrogen standards. As the implanted-hydrogen standards are the most popular, the authors in [46] have undertaken a study to determine the thermal- and radiation stability of these standards. It has been shown, particularly, that out-diffusion of hydrogen implanted in crystalline silicon occurs after annealing at 100 °C, while hydrogen standards prepared by implantation into heavy ion-bombardment amorphized silicon targets do not show out-diffusion after annealing at 295 °C.

Samples of pyrolytic carbon, CH_x, from pyrolysis of methane and amorphous silicon, SiH_x, from plasma deposition of silane, having constant hydrogen distributions, are shown [47] to be stable during ion bombardment and vacuum storage and are therefore well suited as calibration standards for hydrogen profiling by resonance techniques. They must, however, be calibrated from samples with known hydrogen content.

Another problem in hydrogen profile studies by NRRA is that it requires a sample placed in vacuum of $< 10^{-6}$ Torr. This can cause various changes because of the high mobility of hydrogen in many materials. To avoid this difficulty to some extent *Horn* et al. [48] utilized a differential pumping system to create a small gas chamber which can be pressurized to 1 Torr, while maintaining 10^{-6} Torr in the accelerator-beam line. The gas pressure in a gas cell in this non-vacuum environment is limited by pumping speed.

a)   The $^1H(^{15}N, \alpha\gamma)^{12}C$ Reaction

This reaction proposed by *Lanford* et al. [49] is widely used as the most suitable resonance method of hydrogen profiling to study the near-surface region of materials. Hence, it will be discussed in detail in this section. Advantages of this reaction for hydrogen analysis include the following:

1)   any material can be analysed
2)   the detection limit is rather low
3)   the depth resolution is superior
4)   the depth range is comparatively large
5)   the measurement procedure is rather simple

To use this reaction for hydrogen profiling, the surface of a sample is bombarded by accelerated ^{15}N, ions producing α particles and the residual nucleus

^{12}C. If the latter occurs in an excited state, the yield of characteristic 4.43 MeV γ rays, emitted due to the de-excitation of this level to the nuclear ground state, will be proportional to the hydrogen content in a sample. Thus, the hydrogen presence in a sample may be observed by detecting ^{4}He particles or γ rays. The latter is more convenient, as a detector can be placed outside a vacuum chamber, and it is not necessary to prevent the scattering of ^{15}N ions in the detector. However, the background γ radiation from natural and cosmic sources, as well as impurity-induced reactions with γ ray emission should be taken into account.

The excitation function of the reaction ^{1}H(^{15}N, $\alpha\gamma$)^{12}C in the ^{15}N ion energy range 0–14 MeV (Fig. 5.1 [50]) is characterized by two well-separated narrow resonances. The strong and narrow resonance at 6.4 MeV seems to be preferable for hydrogen analysis as it limits the size of the accelerators required for the experiments, and at this resonance energy the ^{15}N stopping power is higher. Earlier publications on the ^{15}N hydrogen profiling method adopted the literature values, $\Gamma_1 = 13.5$ keV and $\Gamma_2 = 19$ keV, for the width of the first and the second resonances, respectively. However, soon after accelerated ^{15}N ions started to be used for hydrogen determination, it was found that the resonance width $\Gamma_1 = 13.5$ keV appeared to be too large to explain the experimental peak width for surface-hydrogen contamination of samples. *Lanford* [51] found $\Gamma_1 = 6$ keV. Width values of this resonance were later re-estimated several times. The latest value $\Gamma_1 = 1.8 \pm 0.45$ keV was obtained by *Maurel* and *Amsel* [19] using high-precision resonance-width measurement techniques; it has also been confirmed by *Damjantschitsch* et al. [52].

The typical experimental scheme for hydrogen profiling using ^{15}N techniques (and other techniques where high-energy γ rays are counted) is shown in Fig. 5.2 [47]. All parts of the beam tube and target chamber are constructed for UHV

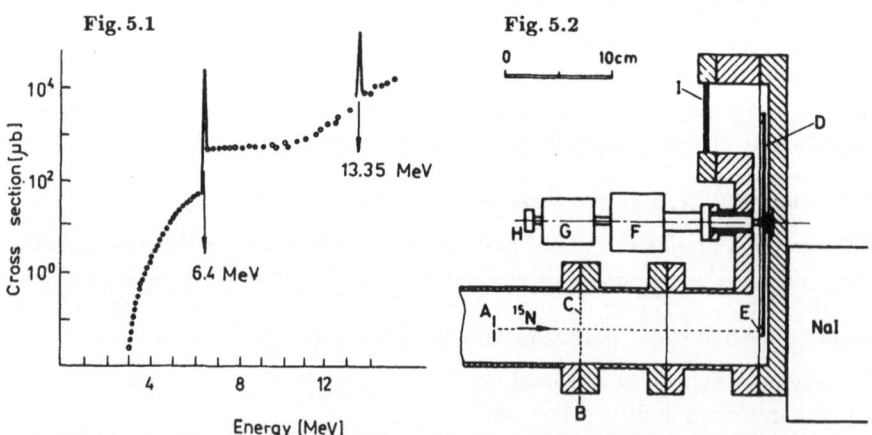

Fig. 5.1

Fig. 5.2

Fig. 5.1. Cross-section of the reaction ^{1}H(^{15}N, $\alpha\gamma$)^{12}C as a function of energy of the ^{15}N ions

Fig. 5.2. Experimental chamber for resonance depth-profiling of hydrogen by γ ray detection. A — beam defining collimator, B — insulator, C — grid for suppression of electrons, D — sample holder, E — sample, F — rotary feed-through, G — step-motor, H — position indicating potentiometer, I — inspection window

operation. The γ ray detector, usually NaI, is placed directly behind the target to maximize the detector solid angle. The entire target chamber is electrically isolated and acts as a Faraday cup. To minimize the errors in beam-current integration, a grid for secondary electron suppression is used. In the case of samples made of insulating material, the surface is coated with a graphite film to avoid charge build-up. Since the technique demands a change of accelerator energy for each data point, many specimens (e.g. 30 [47]) can be remotely rotated into the ion beam.

Figure 5.3 [9] shows two spectra measured with an NaI detector: one with the beam on and the other with the beam off. The difference between the two spectra lies in the 4.43 MeV γ ray peak and its associated escape peaks as well as in the continuous amplitude-distribution due to the Compton effect. The raw data on such measurements are given by the number of counts observed in a single-channel-analyser window set in a particular spectrum region (see arrows in Fig. 5.3). Since the γ rays from this reaction are characterized by rather high energy, the background in this energy region is relatively small.

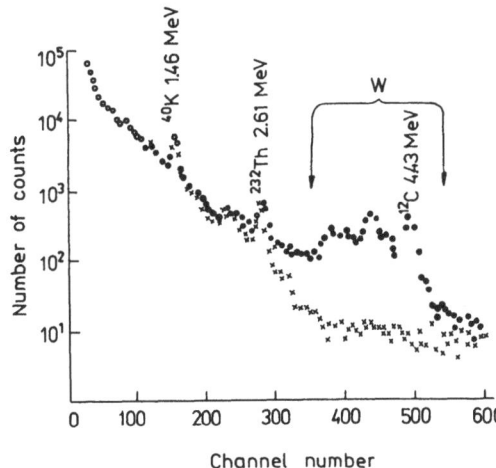

Fig. 5.3. Total ($\bullet$) and background ($\times$) γ ray spectra accumulated with a NaI detector from a hydrogen-implanted silicon target. W designates the window for the recorded region of the ^{1}H(^{15}N, $\alpha\gamma$)^{12}C reaction yield

As unique characteristics of an excitation function for the reaction ^{1}H(^{15}N, $\alpha\gamma$)^{12}C have been obtained, there arises a question concerning the depth resolution limit, the maximum possible depth of the analysis and the detection limit for this hydrogen profiling technique.

Various factors affecting the resolution were examined in detail by *Amsel* and *Maurel* [18]. The limiting value for depth resolution was shown to be conditioned, not only by the resonance width and by the energy distribution of an ion beam itself, but also by Doppler broadening. ^{15}N ions are accelerated by a tandem accelerator, which allows the exact fixation of their energy, with an energy spread of about 2 keV [23]. The contribution to ΔE_0 due to Doppler broadening at room temperature is $\sim$5 keV [18]. Using (3.20) and (3.19) we obtain, for normal incidence, $\Delta x \approx 40$ Å for the surface region ($\Delta E_s = 0$) of a silicon sample ($S_1 = 1.5$ MeV/μm, as deduced by *Ziegler* et al. [9]).

35

This very low value of Δx suggests a clean surface for the samples being analysed, as straggling in the hydrocarbon contamination layer on the sample surface (similar to the Doppler effect) reduces the advantage of the small value of Γ. The Doppler broadening may, however, be beneficially used for directly measuring the vibrational-speed distribution of hydrogen atoms absorbed on clean surfaces [20] or bonded in different gases [48].

Figure 5.4 [52] shows the yield of this reaction vs ^{15}N ion-energy for two hydrogen-free bulk iron samples. One of the samples is coated by an evaporated gold layer 1000 Å thick. The measurements extend beyond the second resonance at 13.35 MeV. The surfaces of both samples are covered with layers of an H-containing contaminant, including the interfacial layer between Au and Fe. Due to this, the two surface hydrogen peaks (at the first and second resonance energies) appear in an excitation curve for the Fe + Au sample. As can be seen from Fig. 5.4, hydrogen profiling can be performed up to the energy of the second resonance. However, complications due to the superimposed contributions of the on-resonance and off-resonance yields already occur at about 10 MeV. This energy corresponds to less than 1 μm of profiling depth and corrections are necessary for profiling beyond this equivalent depth. Hydrogen profiling in silicon can be performed up to x_{max} of about 4 μm.

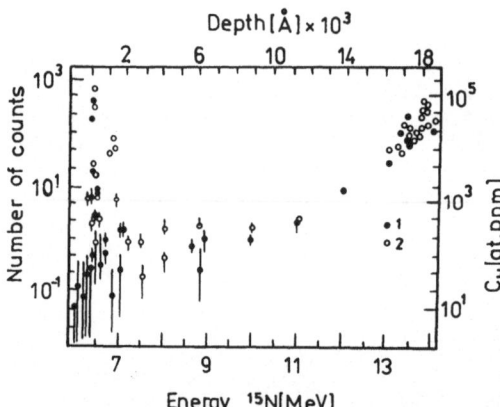

Fig. 5.4. Yields of the ^{1}H(^{15}N, $\alpha\gamma$) ^{12}C reaction from a Fe sample (•) and a Fe sample coated by gold film (○), as a function of the ^{15}N ion energy

The minimum hydrogen concentration level that may be measured by the given method depends mainly on the background level of natural and cosmic radiation, shown by crosses in Fig. 5.3. *Damjantschitsch* and co-workers [52] describe an experimental device built in order to improve significantly the sensitivity of this method. A multiple shield system has been integrated into the beam line of the tandem accelerator and the background level, as seen by the NaI detector, was reduced by a factor of about 100 relative to a usual non-shielded spectrometer ($\sim$0.1 at.% [45, 53]). As for the contribution of the impurity-induced reactions, it should be taken into account only when deuterium and lithium are present in a sample. Reactions with other elements are suppressed by the high Coulomb barrier.

It should be noted that, in order to achieve this low detection limit, it is necessary to provide vacuum conditions and surface treatment such that the surface peak could be reduced considerably. The contribution of the non-resonant interaction with surface hydrogen worsens the analysis sensitivity, both in the close vicinity to the surface and in the deeper bulk. It is important to have a very clean vacuum in the target chamber, since otherwise a gradual build-up of hydrocarbons on the sample surface will cause the depth scale and sensitivity, as well as accuracy of the absolute hydrogen depth profiling, to vary during measurement. It was shown [47] that a vacuum in the region of 10^{-9} Torr is necessary for stable conditions.

Amsel et al. [20] remarked that another advantage of the very narrow resonance width of this reaction is an essential decrease with respect to, say, the $^1H(^{19}F, \alpha\gamma)^{16}O$ reaction of the background to measurements below the surface originating from the surface contamination peak. As the tails of isolated resonance are Lorentzian, the narrower Γ results in a lower background contribution at a given depth.

The above technique of hydrogen profiling based on ^{15}N ions has been very successful in determining the concentration vs depth, but has been unable to measure the lateral distributions of hydrogen. Because of the wide range of potential applications for a method which is capable of measuring the three-dimensional distribution of hydrogen in solids, *Lanford* and *Burman* [54] studied the possibility to do this using the well-established ^{15}N hydrogen profiling technique.

Well known ion-beam-analysis techniques were extended to measure lateral distributions by focusing the incident ion beam to a microbeam (typically of order 1–10 μm) and rastering this beam across the sample, but these appear not to work reliably for hydrogen; due to the high mobility of hydrogen, the high-energy density in the microbeam drives the hydrogen away from the analysis point.

The extension of *Lanford* and *Burman* makes use of the bombardment of the sample with a large uniform ^{15}N beam and detects the ^4He particles emitted in the $^1H(^{15}N, \alpha\gamma)^{12}C$ nuclear reaction to determine the lateral distribution of hydrogen within the plane parallel to the surface of the sample at the depth where $E_1 = E_R$. This procedure is shown schematically in Fig. 5.5. The experiment

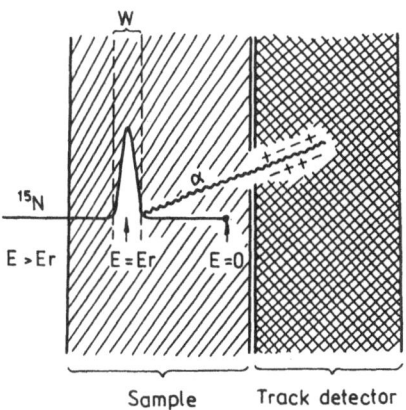

Sample   Track detector

Fig. 5.5. Autoradiographic technique to measure lateral distribution of hydrogen by means of the $^1H(^{15}N, \alpha\gamma)^{12}C$ reaction; W designates the hydrogen detection resonance window

requires the sample to be in the form of a "thin section" with a thickness 10–20 μm. At the depth where $E_1 = E_R$, nuclear reactions will occur with the emission of both γ rays and α particles. The α particles emitted in the direction of the ^{15}N beam will have an energy of 4.25 MeV and a much longer range than the residual ones of the 6.4 MeV ^{15}N ions.

Thus, by choosing the thickness of the thin section appropriately or by using an absorber foil behind the sample, ^{15}N ions from the incident beam can be ranged out, allowing the α particles to emerge from the back of the sample. The ^{4}He particles can be detected at 0°, with a low background count, and used to form an autoradiograph by placing a plastic track detector in contact with the back of the sample. By varying the beam energy and recording the autoradiographs at each depth, the full three-dimensional distribution of hydrogen within the sample is determined.

The method was checked on a test sample that was prepared by ion implantation of hydrogen through a wire grid into a thin silicon crystal. The lateral resolution can be estimated to be approximately 15 μm and is limited by the spread resulting from α particles emitted at various angles. The measurements are complicated by the background in the autoradiographs due to the nuclear reaction ^{15}N + ^{1}H $\rightarrow$ ^{12}C + ^{4}He (with ^{12}C in its ground state) on the surface hydrogen.

The passive track detector could be replaced by an electronic two-dimensional position-sensitive detector. Such an approach would offer the great advantage of a live time-display and would eliminate the need to replace the track detector for every exposure.

b) The ^{1}H(^{19}F, $\alpha\gamma$) ^{16}O Reaction

This reaction was first introduced for the study of hydrogen depth profiling using nuclear resonance techniques by *Leich* and *Tombrello* [55]. The interpretative problems have been examined in [56]. This reaction exhibits a strong isolated resonance at a ^{19}F energy of 16.44 MeV, with a peak cross-section of 0.5 b and fwhm of about 90 keV and a weaker one at a 6.42 MeV beam energy ($\sigma_R \approx$ 0.1 b; $\Gamma \approx$ 45 keV), producing 6.1, 6.9 and 7.1 MeV γ rays from the de-excitation of the residual ^{16}O excited nuclei.

Usually the high-energy resonance is chosen for hydrogen depth profiling. The depth resolution of this method is determined mainly by a resonance width which varies from 250 Å at the surface to 300 Å at a 4000 Å depth in silicon [56]. This depth is the maximum analysable one and is limited by the presence of another resonance at 17.56 MeV (the profiling depth may be increased up to 1.2 μm by using the resonance at 6.42 MeV). The sensitivity of the ^{19}F method was estimated to be $(2–3) \times 10^{-3}$ at.H/at.Si (see e.g. [45]). However, care must be taken to ascertain that γ rays from other competing reactions are not being detected, since the energy of ^{19}F ions must be higher than 16 MeV, which is sufficient energy to overcome the Coulomb barrier for the elements of $Z \leq 5$. The non-resonance interaction of ions with the surface hydrogen contamination complicates the determination of small hydrogen concentrations in the bulk of the sample. Another factor which limits the sensitivity appears to be a contribution of γ rays from the interaction of ^{19}F ions with hydrogen inside the sample at energies below the main resonance (16.44 MeV). This is particularly true for thick samples with uniform hydrogen concentration, where this contribution may make up one-quarter of the total yield [56].

Clark et al. [56] and *Xiong* et al. [57] have compared hydrogen profiling by ^{19}F ions with that by ^{15}N ions. In Fig. 5.6 [57], four peaks obtained by profiling the hydrogen contamination on the surface of the Ta foil are compared in terms of surface depth resolution. It has been found that the widths (fwhm) of the surface peaks are 14, 48, 34 and 90 keV for the ^{15}N first resonance, ^{19}F first resonance, ^{15}N second resonance and ^{19}F second resonance techniques, respectively. These values correspond to those of a depth resolution in Si(nm) of 9.6, 23.6, 24.5 and 46.2, respectively, which are typical numbers obtained without special precautions. Therefore, they cannot serve as the best characteristics of the methods, but may be used only for comparison.

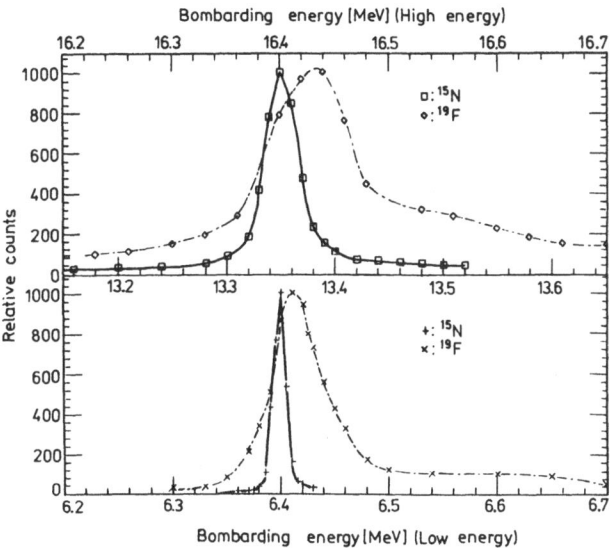

Fig. 5.6. Hydrogen depth profiles of a surface contamination layer on a Ta foil obtained by the four NRRA techniques: ^{15}N 13.35 MeV ($\square$), ^{19}F 16.44 MeV ($\diamond$), ^{15}N 6.4 MeV (+) and ^{19}F 6.42 MeV ($*$)

The ^{1}H(^{15}N, $\alpha\gamma$) ^{12}C reaction seems to be a more successful resonance method, which provides an excellent depth resolution and high sensitivity, using both the resonance at 6.4 MeV and at 13.35 MeV. Nevertheless, the use of the ^{1}H(^{19}F, $\alpha\gamma$) ^{16}O has the advantage of a higher resonance cross-section and width compared to the ^{15}N method. Because of this, the analysis is more rapid and, thus, is characterized by a lower value for the total damage introduced into a sample while obtaining a profile. Furthermore, ^{19}F is an easier beam to obtain from a tandem accelerator than that of ^{15}N ions.

The ^{19}F method, however, is characterized by a lower sensitivity in spite of the smaller underlying background in the γ ray energy spectrum. This may be explained both by a large cross-section in a non-resonance area and by a high probability of γ ray emergence from impurity elements, caused by high energy ions overcoming the Coulomb barrier. Moreover, because of the larger resonance width and smaller distance between resonances, the method considered seems not to be advantageous over the method employing ^{15}N ions, either in depth resolution or in maximum analysable depth.

c) The ^{1}H(^{7}Li, γ) ^{8}Be Reaction

This reaction [45, 58] has a resonance in the excitation function for bombarding lithium ions of 3.07 MeV, producing 17.7 and 14.7 MeV γ rays. The fwhm of the resonance and its cross-section were re-estimated in [45] to be 81 keV and 4.8 mb, respectively. It has been used [9, 45] for concentration profiling of hydrogen implanted in silicon. The reaction may be applied to hydrogen-concentration distribution measurements, up to the depth of 6–8 μm (silicon) with a detection limit of $\sim$0.1 at.%. The depth resolution for silicon, however, is about 0.1 μm. Thus, with regard to the depth resolution this method is inferior to other resonance techniques, but it is characterized by the highest analysable depth.

The second resonance ($E_R = 7.11$ MeV, $\Gamma = 1.2$ MeV, $\sigma_R = 135$ mb) seems to be more adapted to the global determination of hydrogen in solids, with the detection limit of about 10^{-3} at.H/at.M [45].

d) The ^{1}H(^{11}B, α) ^{8}Be Reaction

A cross-section of about 8 mb/sr (isotropic) at the ^{11}B ion resonance energy of 1.793 MeV is characteristic for the ^{1}H(^{11}B, α) ^{8}Be (^{8}Be $\rightarrow \alpha + \alpha$) reaction, which results in the production of three α particles [9, 59].

The experimental scheme and the excitation curve of the ^{1}H(^{11}B, α) ^{8}Be nuclear reaction from the hydrogen-implanted silicon sample is shown in Fig. 5.7 [9]. The sample is tilted at 45° to the incident ^{11}B beam direction, and α particles

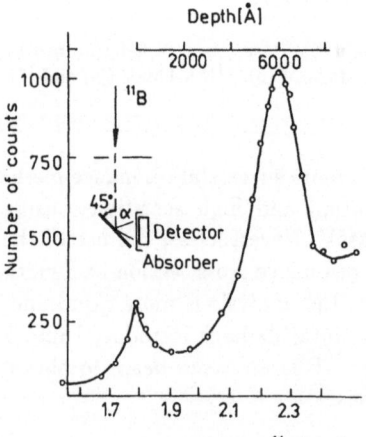

Fig. 5.7. Excitation curve of the ^{1}H(^{11}B, α) ^{8}Be reaction from a hydrogen-implanted Si target

with an energy $E < 4\,\mathrm{MeV}$ are detected in the reflection geometry by a surface-barrier detector covered by a thin absorber to eliminate scattered [11]B ions. The number of counts is proportional to the hydrogen content in a given sample. The solid line is the result of a convolution procedure, assuming that the hydrogen profile is a product of both a thin layer of surface contamination and a hydrogen distribution at the implantation depth of about 6100 Å. The best fit for the excitation curve was obtained by taking into account several Gaussian distributions with the total standard deviation $s_t^2 = s_R^2 + s_H^2 + s_s^2$ where $s_R = 350\,\text{Å}$ is the standard deviation corresponding to the resonance width $\Gamma = 66\,\text{keV}$; $s_H = 620\,\text{Å}$ corresponds to the implanted-hydrogen profile and s_s is the standard deviation due to the straggling of the [11]B analysing beam. The final term has not been found to be important in this measurement. i.e. the depth resolution is limited by the resonance width.

The detection limit of the [11]B method cited in [45] is $\sim 10^{-2}\,\text{at.H/at.Si}$. However, the maximum analysable depth mentioned is $0.4\,\mu\mathrm{m}$, while in Fig. 5.7 the depth profiling is shown to be obtained up to $0.8\,\mu\mathrm{m}$. Also, the value of depth resolution, 500 Å, published in [45,59] is in contradiction with that of $\sim 800\,\text{Å}$ which can be calculated from the known resonance width $\Gamma = 66\,\text{keV}$ and the measured stopping power of $810\,\text{keV}/\mu\mathrm{m}$ for a boron beam in Si [9].

e) The $^1\mathrm{H}(^{18}\mathrm{O},\alpha)\,^{15}\mathrm{N}$ Reaction

The first experimental measurements of the resonant reaction $^1\mathrm{H}(^{18}\mathrm{O},\alpha)\,^{15}\mathrm{N}$ ($E_R = 11.25\,\text{MeV}$; $\Gamma = 43\,\text{keV}$; $\sigma_R = 75\,\text{mb}$) has been recently described by *Trocellier* et al. [45]. This reaction has the advantage of producing α particles ($E < 3.4\,\text{MeV}$) in an energy region with a low background. Therefore, the sensitivity of hydrogen determination is estimated to be rather high, i.e. about $10^{-4}\,\text{at.H/at.Si}$. Since other analytical characteristics (depth resolution = 200 Å and the maximum analysable depth = $3.5\,\mu\mathrm{m}$) appear to be satisfactory, the reaction can be used as a tool for hydrogen depth profiling in solids.

f) The $^1\mathrm{H}(^{13}\mathrm{C},\gamma)\,^{14}\mathrm{N}$ Reaction

Since this reaction requires a very high incident energy due to the fact that $E_R = 22.55\,\text{MeV}$, its analytical capabilities were never examined [45].

g) Examples of Applications

The resonance nuclear reactions can be used for hydrogen analysis and depth profiling over a wide range of scientific and technological problems. Let us consider now some of the most interesting applications.

The reaction $^1\mathrm{H}(^{15}\mathrm{N},\alpha\gamma)\,^{12}\mathrm{C}$ has been used by *Lanford* [51] to study the temperature dependence of the superconductive transition T_c in $\mathrm{Nb_3Ge}$ vs the content of hydrogen in a sample as introduced by etching with HF. Correlation between hydrogen concentration in a sample and T_c has been found. It is of interest to note that even superconductive samples that were not etched in HF had an appreciable hydrogen concentration (about 1 at.%). From this arises the question whether hydrogen is an important factor that affects T_c or only a correlated parameter of some other problem? To answer the question, much work needs to be done in this interesting and important field. Now the phenomenon of high temperature superconductivity has been discovered, the study of the effect of hydrogen on superconductive material properties may be very useful.

The capability to contain ultra-cold neutrons (UCN) in a neutron "bottle" made of various materials seems to be one of the central problems in experiments with UCN. Neutrons with an energy of $\sim 10^{-7}$ eV undergo total reflection from the walls of many materials. This total internal reflection leads to the possibility of containing neutrons in bottles.

Nevertheless, there is a finite probability that a neutron will be lost due to nuclear capture or due to inelastic scattering during reflection. The possibility of escaping from the bottle depends on a potential step, whose height is determined by the nucleus of which the solid is composed.

The predicted neutron loss per reflection happened to be two orders of magnitude less than the experimental determination for graphite or glass. One possible explanation of this discrepancy is that hydrogen is present near the surface of the materials used to make neutron bottles. Surface hydrogen lowers the potential barrier. Hence, neutrons can penetrate further, and remain inside the solid longer, during a reflection. This results in the decrease of neutron losses. In addition, hydrogen has a large cross-section for interactions with neutrons.

The measurements of hydrogen concentration profiles in materials used for UCN bottles by the ^{15}N ion technique [51] showed rather high hydrogen concentration. If one now includes these measured hydrogen profiles in a calculation of the probability of UCN loss, one finds qualitative agreement between calculation and experiment. This has led to the usage of materials which, not only have the correct nuclear properties to make a good UCN bottle, but also have a surface which would not attract a hydrogen-rich film. Amorphous materials have proved to be the most "clean" relative to surface hydrogen.

One of the most interesting applications of the ^{15}N and ^{19}F ion methods [51, 56] appears to be the study of hydrogen behaviour in amorphous silicon. The latter is considered to be a promising material for making inexpensive large-area solar cells. If a large quantity of hydrogen (15–35 %) is introduced into amorphous silicon, it turns into a semiconductor. Analysis of hydrogen-concentration profiles may play a key role in the study of this phenomenon. The correlation between hydrogen content in amorphous silicon nitride and the etching rate can also be determined by this method. The etching rate is important for the fabrication of microelectronic circuits.

The introduction of small amounts of water into the quartz structure has a profound influence on the mechanical properties of this material. Therefore, *Klark* et al. [56] used the reaction ^{1}H(^{19}F, $\alpha\gamma$) ^{16}O to determine the depth distribution of hydrogen in several quartz samples. All samples examined exhibited the complex profile which consisted of three regions: (1) a narrow hydrogen surface contamination region; (2) a broad region below the surface of approximate thickness 2000 Å; and (3) a region deep in the crystal where observed yield is partially due to off-resonance nuclear reactions occurring with hydrogen in the surface and near-surface regions.

To clarify the role of hydrogen in electro-deposited hard gold, extensively used for sliding electrical contacts, the hydrogen depth profiles in gold films were examined using the reaction $^1H(^{19}F, \alpha\gamma)\,^{16}O$ [56]. Profiles measured indicated that hydrogen is uniformly distributed throughout the film. However, the samples plated with a bath chemistry containing cobalt ions were shown to have surprising amounts of hydrogen (up to 9 at.%). The samples produced in cobalt-free bath chemistry contain 0–3 at.% of hydrogen.

Another application of the resonance method based on the reaction $^1H(^{15}N, \alpha\gamma)\,^{12}C$ is the study of the glass hydration process by measuring the thickness of the hydrated layer on the surface of the object. Glass hydration is both of physical interest itself and important from the point of view of its wide spread applications. For example, it may help in the understanding of the reasons for glass fractures at stresses of about 0.001 of its theoretical strength and the hydrofracture of minerals, as well as to date objects and even to study earthquakes. Figure 5.8 [51] shows a series of hydrogen profiles in a glass sample, measured with the help of ^{15}N ions at different times beginning from the hydration onset.

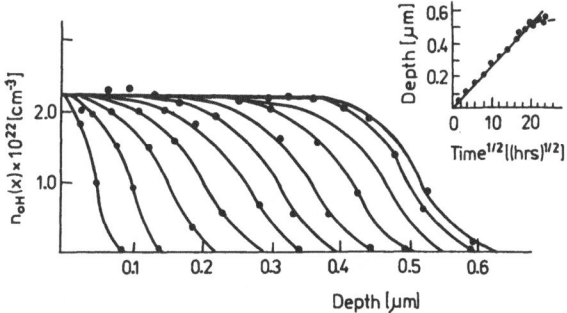

Fig. 5.8. The measured hydrogen profiles of a glass sample hydrated for varying length of time, and the thickness of the hydration layer vs square root of the hydration time

Hydration was carried out in distilled water at 90 °C. The hydrated layer thickness (depth to 1/2 of C_{max}) is shown in the right upper corner of Fig. 5.8. The square root dependence illustrates the fact that the hydration process is diffusion limited. The slow-down of the diffusion rate to the rate of the surface dissolution results in the steady-state condition of a hydrated layer thickness that is constant with time.

The comparison of hydrogen profiles measured by this reaction with those of sodium obtained using the $P + ^{23}Na \rightarrow ^{24}Mg + \gamma$ ray reaction (resonance energy 402 keV) [51, 60] and backscattering spectra [61] clearly shows considerable enlargement of the sodium step region. Apparently, this means that the hydration process in soda-lime glasses involves an interdiffusion mechanism based on the replacement of Na^+ ions with H_2O^+ ions.

This slow alteration of the surface layer may occur as a result of the reaction between glass and its environment. Since it is probably universal, always beginning with the manufacture of the object and never ending, the analysis of the surface composition has potential importance for the study of archeological and historical artifacts. A basic understanding of the glass hydration mechanisms involved may lead to new and better conservation practices for the preservation of artifacts. Due to the lack of accessibility to artifacts for analysis, *Lanford* [61] even suggests the installation of an accelerator in the Louvre!

The very interesting phenomenon of the gettering of hydrogen caused by Ti ion-implantation of Fe, has been examined by *Rauch* [60]. Similar results were found for Ni targets implanted by Ti ions except that the amount of gettered hydrogen was smaller. The Fe samples were implanted by different fluencies of Ti ions and the ^{15}N method was used for H profiling. For Ti profiling the resonance reaction ^{48}Ti(p, γ) ^{49}V was employed.

It has been observed that a Ti-containing surface layer is generated, which grows in thickness with increasing Ti fluency. It is interesting that the Ti penetration into the samples is one order of magnitude deeper than the theoretical range, 11 nm, of 30 keV ions. This layer contains increasing amounts of hydrogen (up to 30 %). Although the Ti profiles reach somewhat deeper, a rough correlation between the shapes of the Ti and H profiles exists. The study of this effect indicates that the hydrogen is taken up from the residual gas of the vacuum chamber. The author *Rauch* [60] believes that free H atoms are formed, which can enter the sample by dissociation of hydrocarbon and water molecules at the sample surface. This effect is chemically activated by the impinging ions during the Ti implantation.

The theory that radiation damage was the cause of the hydrogen gettering could be excluded since samples implanted by Fe and Ni did not contain hydrogen. This supports the view that the strong hydrogen affinity of the implanted element is responsible for the gettering. Therefore, in ion-implantation studies with ions of hydride-forming elements, like Ti, Zr, Sc and Y, one should expect that gettering hydrogen might be present in the surface layer.

In the annealing measurements it was found that the gettered hydrogen is retained in the samples up to surprisingly high temperatures and it is released completely in a narrow temperature interval. This means that the gettering of hydrogen caused by ion implantation leads to unusual metal-hydrogen system formation.

New ways to apply resonance methods are being looked for based on studies of steady-state distributions, e.g. [62–66], and hydrogen diffusion processes. The latter will be considered in Chap. 6.

5.2 Energy Analysis of Nuclear Reaction Products

In the case of resonance nuclear reactions, the energy spectrum of reaction products is of no interest for depth profiling. If there are no resonances in the cross-section of a nuclear reaction that varies smoothly with energy, but there are some other peculiarities of the reaction, which allow us to distinguish the events of interaction with hydrogen among the competing processes on the nuclei of the sample matrix, then the depth information stems from an energy analysis of the emitted particles.

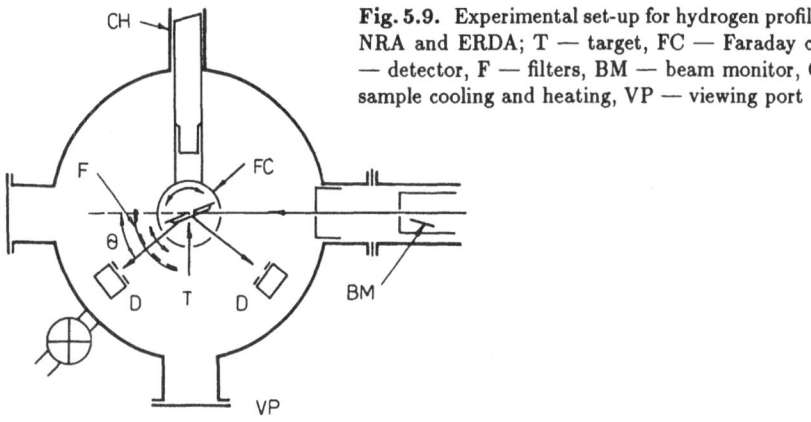

Fig. 5.9. Experimental set-up for hydrogen profiling by NRA and ERDA; T — target, FC — Faraday cup, D — detector, F — filters, BM — beam monitor, CH — sample cooling and heating, VP — viewing port

The schematic representation of a typical experimental setup for the hydrogen depth profiling by nuclear reaction analysis, as well as by nuclear recoil detection, is shown in Fig. 5.9 [67]. A beam of collimated projectiles from a Van-de-Graaff or a tandem accelerator with energies from one-tenth to tens of MeV strikes the target, which may be tilted at various angles to the beam direction. The energy distribution of emitted charged particles may be measured by a surface-barrier silicon detector, in a transmission as well as in a reflection geometry, at various θ values with respect to the incident beam. All beam particles scattered from the target may be stopped with filter foils placed in front of the detector. The resulting neutron energy spectrum is measured via neutron time-of-flight from the target to an organic scintillator several meters away from the scattering chamber. This requires a sample bombardment of nanosecond bursts of accelerated particles. The temperature of the sample may be stabilized by means of its cooling or heating. The beam current is monitored using a Faraday cup or a special monitor target introduced periodically into the beam.

The energy of the emitted particle is sensitive to both the kinetic energy of the incident beam and the angle of particle emission as well as to the reaction Q value. Furthermore, the probing beam continuously loses energy on penetrating the sample as does the emitted particle, if it is charged, on its way out of the sample. As a consequence, the measured yield spectrum according to (2.1) and (2.5–7) varies considerably in shape, not only as a function of incident beam energy and reaction excitation function, but also with the angle of the detector and of the sample surface to the beam. Nevertheless, by analysis of the emitted particle spectrum the hydrogen concentration profile can be unambiguously obtained using (2.3).

Multiple scattering of the analysing particles by a sample material results in an angular spread which increases at small energies, i.e. at the end of the particle range. As all of the reactions to be discussed are characterized by a strong dependence of the emitted particle energy on the angle, the condition $R\sin\psi > x_{\max}$ should be fulfilled to avoid errors in the depth profiling due to the corresponding resolution deterioration [see (3.9, 10)]. This means that the average incident particle range should be greater than the maximum depth of the region to be analysed. Moreover, this condition should be satisfied since radiation damage is produced primarily near the end of the range of the probing ions.

The non-resonance method has a principal advantage relative to the resonance method: it allows us to perform the analysis throughout the whole sample depth in a single measurement without changing the incident particle energy.

a) The $^2\text{H}(^3\text{He}, \text{p})\,^4\text{He}$ Reaction

The cross-section for the $\text{D}(^3\text{He}, \text{p})\,^4\text{He}$ reaction varies widely with energy and possesses a maximum of approximately 50 mb/sr for 0.65 MeV ^3He particles at a proton laboratory angle $\theta = 165°$ [29]. It is not practical, however, to use this reaction for resonance profiling, because the peak in the excitation function is very broad [68]. This reaction is highly exothermal ($Q = +18.35\,\text{MeV}$), which is why the incident ^3He ion energy, $E_1 = 0.7\,\text{MeV}$, results in protons and α particles of energies $\sim 13\,\text{MeV}$ and $\sim 2\,\text{MeV}$, respectively, emitted at a backward laboratory angle $\theta = 165°$. This reaction, described in detail by *Dieumegard* et al. [29, 69], allows us to determine the heavy hydrogen isotope deuterium. The reaction is more suitable than other nuclear physics methods of deuterium profiling in the near-surface region of solids. The behaviour of hydrogen and deuterium in materials differs only in some isotopic effects. Hence, if a sample matrix contains either its own hydrogen or is contaminated by it, it is reasonable to replace the hydrogen by deuterium to study the hydrogen-involving phenomenon. The analysis is carried out with the help of a low-energy accelerator (in order to accelerate $^3\text{He}^{2+}$ ions up to 700 keV, one should use a voltage of 350 kV). The angle should be the maximum possible, as it corresponds to the minimum flow of backscattered ^3He ions and to the minimum derivative $dE_3/d\theta_1$. The latter reduces the effects of the angular spread and means that there is no need for a precise angular setting.

The differential cross-section shows a weak monotonic dependence on the particle emergence angle. Its energy dependence [important for the concentration calculation by (2.3)] may be described by the analytical approximation [68]

$$d\sigma/d\Omega = 475 E_1^3 \Big/ \left[1 - 26.2 E_1^{3.43} + 36.5 E_1^{3.91} \right] \ , \tag{5.2}$$

where E_1 is measured in [MeV] and the cross-section in [mb/sr].

The analysis can be performed by detecting the energy spectrum of α particles as well as that of protons. *Dieumegard* et al. [29] compared these two variants and showed that detection of protons presented distinct advantages with respect to the more conventional method of detecting the associated α particles. These advantages concern depth resolution, the background contribution due to the reaction with impurity light elements, etc.

An interesting feature of the reaction $\text{D}(^3\text{He}, \text{p})\,^4\text{He}$ is that, at backward angles, the energy of the particles emitted at a given angle decreases when the bombarding ^3He ion energy increases (the derivative $\partial E_3/\partial E_1 < 0$). Then, in the expression (3.8), the terms in brackets will be opposite in sign, i.e. the stopping of a detected particle (at S_1 comparable to S_2) does not improve the depth resolution. In the case of high-energy proton detection $S_2 \ll S_1$, the expression (3.8') may be presented as follows

$$\Delta x \approx -\Delta E \left(\frac{\partial E_3}{\partial E_1} \frac{S_1}{\sin \psi} \right)^{-1} . \tag{5.3}$$

Therefore, in spite of the fact that the α particle detector has better resolution compared to a proton detector, the registration of protons leads to better resolution.

The depth resolution of deuterium profiling with the $D(^3He, p)\,^4He$ reaction was examined in detail by *Dieumegard* et al. [29]. The best resolution of 200 Å was reached in the region near to the surface of a silicon sample, without the stopper foil, by tilting the target with respect to the beam ($\psi = 20°$), whereas normal incidence results in a surface resolution of 500 Å. It should be mentioned that, in this case, the 13 MeV protons were detected using a spectrometer based on a high cost 1500 μm deep silicon surface barrier detector with an overall resolution of $\sim$17 keV. As can be seen from Fig. 3.2, the total resolution shows strong depth dependence. The maximum profiling depth in silicon is no more than $x_{max} = 1 \mu$m (at $\psi = 90°$).

In order to increase the analysable depth, *Giles* and *Wilson* [70] performed deuterium depth profiling using the $D(^3He, p)\,^4He$ reaction in conjunction with bevelled samples and adjustable slit collimation. The production of very small angle bevels, together with a narrow beam probe, permits the attainment of a depth resolution of 0.4 μm at depths as great as 100 μm for stainless steel samples. However, after charging with deuterium each sample was submitted to a polishing procedure (about three hours) that could cause a change in its deuterium content. Besides, the complete deuterium concentration profile of a given sample was obtained as the result of irradiations at more than 25 different points along the sample surface and required 3–4 hours of accelerator time.

To increase the rapidity of the deuterium analysis, *Altstetter* et al. [69] and *Besenbacher* et al. [71] have determined the concentration profile $C(x)$ from the proton counts as a function of incident ^3He ion energy, $N(E_0)$. Because of a large detector acceptance angle, it provides a rapid determination of the total number of deuterium atoms in a layer for which the ^3He ions have significant reaction cross-sections. In order to obtain $C(x)$, the integral equation

$$N(E_0) = \int C(x)\, \sigma(E_1(E_0, x))\, dx \tag{5.4}$$

has to be solved, which leads, however, to additional interpretative problems. The analysable depth can be varied up to 2 μm by adjusting the energy of the incident ^3He ions.

The detection limit of the $D(^3He, p)\,^4He$ method was estimated to be $\sim$10^3 at.ppm [8], but it depends on experimental set geometry, interferences from Li, B and N target impurities and on the maximum permissible beam current which does not cause deuterium distribution distortion in the sample under study.

On the basis of this reaction, *Ilic* and *Altstetter* [72] proposed a method to obtain a concentration pattern for deuterium in thin films. The target investigated is irradiated by ^{3}He ions, in contact with a solid track detector. Resolution of the lateral distribution is $\sim$1 μm at the detection limit for deuterium, 10^{-6} at.D/at.M.

The second example is to determine the deuterium lateral distribution in graphite diaphragms from a Tokamak device reaction chamber. The French AEC nuclear microprobe scanning system, based on sweeping of the focused ^{3}He ion beam (2–3 μm) across the target by electric deflection, was used for this purpose [73]. The emitted protons were detected by an annular surface barrier detector of 100 μm depletion depth covered with a 200 μm thick tantalum absorber to slow the proton energy down enough ($\sim$7 MeV), so that a well-defined peak characteristic of deuterium could be identified in the spectrum.

The method has been applied to measure deuterium depth profiles in amorphous silicon [69] and stainless steel [68], as well as to study deuterium diffusion [74–76] and trapping by defects [71, 77] in solids.

b) The T(p, n) ^{3}He, D(d, n) ^{3}He, H(t, n) ^{3}He and T(d, n) ^{4}He Reactions

The nuclear reactions between the nuclei of hydrogen isotopes are suitable for analysing the concentration distribution of all three hydrogen isotopes using the same apparatus, i.e. by measuring time-of-flight spectra of emitted neutrons. A Van-de-Graaff accelerator producing 2 ns long bursts of protons, deuterons or tritons with the energies of up to 5 MeV was used for this purpose by *Davis* et al. [78].

As the neutrons do not lose energy in exciting the sample, the expressions (2.6, 7) for a coordinate and (3.8) for resolution may be simplified by excluding the dependence on S_2. The expression (3.8) transforms into (5.3), whereas (2.6, 7) turn into $E = E_3(E_1, m, Q, \theta)$. The concentration profile can be calculated using (2.3), where the detector efficiency $\eta(E)$ for neutrons is less than 1 and should be specially determined.

Figure 5.10 [78] shows the neutron spectra from the reaction H(t, n) ^{3}He, obtained by bombarding TiH$_{1.7}$ and Ti targets 10 μm thick with tritons at an

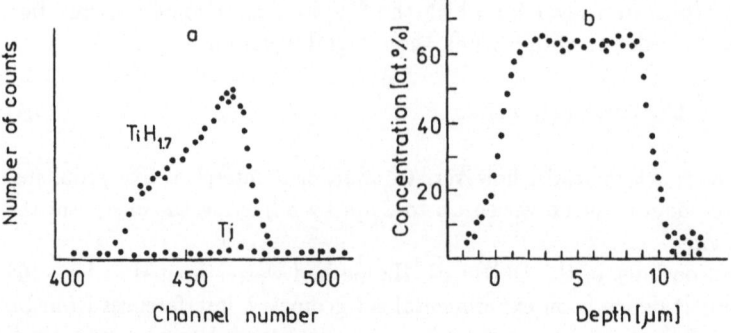

Fig. 5.10. Neutron time-of-flight spectra for 4.5 MeV tritons incident on targets of TiH$_{1.7}$ and Ti (a); depth profile of hydrogen in TiH$_{1.7}$ (b)

energy of 4.5 MeV. The hydrogen concentration profile is given in the lower part of the figure. The essential difference in the shape of these two distributions is due to the fact that the concentration profile was calculated by subtracting the background and removing the effects of the variation of cross-sections and detector efficiency with energy.

The effect of the device resolution is expressed in the smearing of the edges of a uniform hydrogen distribution. *Davis* et al. [78] calculated the depth resolution of hydrogen profiling with the H(t, n) reaction in Ti to be $\sim$0.5 μm, but found it to be limited by the spectrometer time-resolution within 2 μm of the front surface of the sample. Beyond this depth the straggling of the probe beam dominates the overall resolution. The resolution gets worse with depth and reaches $\sim$1 μm at 10 μm. The maximum depth of hydrogen profiling depends on the energy of tritium ions and may reach 50 μm (for titanium), with a resolution at this depth of about 2 μm.

The T(p, n) and D(d, n) nuclear reactions have been applied by *Davis* and *Anderson* [78] to measure the depth profiles of tritium and deuterium in titanium. The technique may be used for titanium sample thickness up to 20 mg/cm^2, achieving a sensitivity of 0.1 at.% and depth resolution of 0.4 mg/cm^2 (0.9 μm). They also show that the T(d, n) ^{4}He reaction can also be applied to determine the depth profile of tritium in materials by the time-of-flight technique.

Such high sensitivity cannot be achieved for the H(t, n) reaction because of the background level due to competing processes that result in the emergence of neutrons. This background is rather high because of contamination of the vacuum system surfaces by hydrocarbons, and because of hydrogen embedding into the collimating slits and apertures of the beam transport systems as a result from accelerator operations with proton beams. Additionally, the reactions (t, n) are usually characterized by a positive Q value and a Coulomb barrier which is comparatively low for light elements, and results in non-negligible reaction probabilities. Neutrons from these background reactions may overlap those from the sample. All of this produces rather low sensitivity for hydrogen determination by the H(t, n) reaction technique; values range for different materials from 5 to 20 at.%.

An interesting technique has been used by *Earwaker* et al. [79] for tritium profiling in solid targets used for neutron production. This was done by studying the zero degree neutron yield in the vicinity of the threshold in the T(p, n) ^{3}He reaction (1.019 MeV). At threshold, neutrons are produced with zero energy in the center-of-mass frame and all appear therefore at 0° in the laboratory frame. As the proton energy is increased, the neutron yield, measured by a detector placed at 0° and subtending a small angle at the source, peaks when all the tritium atoms are exposed to above threshold protons. Thereafter, the yield drops because of the kinematic angular broadening of the neutron beam. The energy width of the rise in neutron yield above the threshold is therefore a measure of the tritium-containing layer.

49

c) The ^{2}H(d, p) ^{3}H Reaction

Figure 5.11, taken from [26], shows the high-energy part of a charged-particle spectrum (a) obtained from 2 MeV deuteron bombardment of a Ni target, and the calculated concentration profile of deuterium (b) implanted in nickel by an ion beam with an energy of 400 keV. Reaction products were detected by a surface barrier silicon detector (with resolution 15 keV) placed at an angle to the direction of incident particles of $\theta = 40°$.

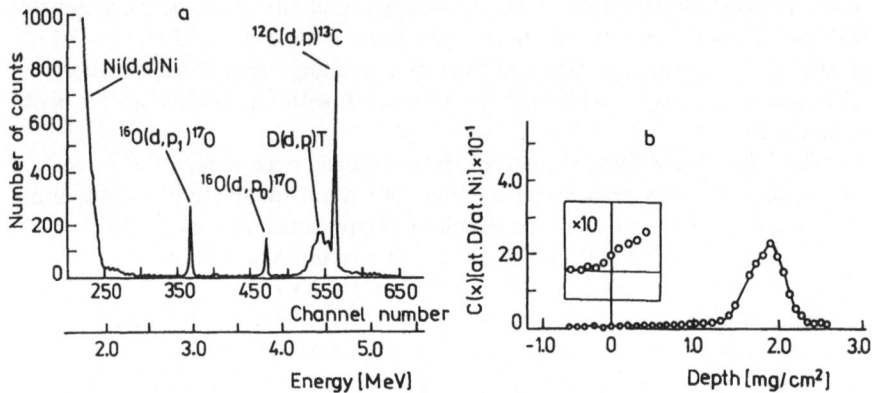

Fig. 5.11. Energy spectrum obtained from 2 MeV deuteron bombardment after implantation of D in Ni (a); computed deuterium depth profile (b)

In the energy spectrum can clearly be seen (i) a group of protons from the reaction D(d, p)T, which corresponds to implanted deuterium; (ii) several peaks due to reactions arising from surface contamination by carbon and oxygen; and (iii) the intense contribution of elastically scattered deuterons. The large flux of scattered deuterons may result in both the distortion of the concentration profile (caused by a pile-up background in the energy spectrum) and a quick destruction of the detector. Since filtering of the elastically scattered deuterons with the help of a retarding foil in front of the detector deteriorates the energy resolution considerably, *Möller* et al. [26] used a simple electrostatic analyser to deflect the deuterons. Also, it can be used to subtract the superimposed proton surface peak from the reaction ^{12}C(d, p) ^{13}C.

In spite of the fact that deuterium implantation has been carried out at a low temperature (160 K), a long tail of the distribution to the surface is observed. The magnification in Fig. 5.11b indicates that a fraction of the distribution exceeds the real surface position. The latter is connected with the technique's resolution of ~0.07 μm near the surface. The increase of the analysis depth up to its maximum (~4 μm) results in the deterioration of the depth resolution to ~0.8 μm due to straggling and multiple scattering. As the protons from the reaction D(d, p)T ($Q = +4.033$ MeV) are characterized by a much lower stopping power compared to the incident deuterons, the expressions (2.6, 7) and (3.8) may be simplified.

50

The detection limit for deuterium profiling using the reaction $D(d,p)T$ is 0.2 at.% (for nickel [26]). It is limited by unavoidable background caused by reactions from surface contamination and within the sample itself. Furthermore, the concentration profile may be additionally falsified by implantation of the probing deuterons into the sample. Reliable data on the concentration profile may be obtained at a content of no less than 1 at.%.

Barit et al. [80] have utilized the $D(d,p)T$ reaction at 1.6 MeV for profiling the deuterium implanted in Ti–T targets by deuteron bombardment during fast neutron generation. It was shown that the substitution of tritium by implanted deuterium causes the neutron yield to decrease in time.

It seems very surprising that there is a discordance in the evaluation of the relative depth resolution for the maximum analysable depth in the reactions $D(d,p)T$ [26] and $H(t,n)\,^3He$ [78] of 20% and 4%, respectively. The reason for this probably lies, not in the difference in type and energy of the analysis ions, but rather in the way in which the resolution is calculated. For example, in [78] the contribution of multiple scattering to the resolution was not taken into account, although according to [26] it is quite significant.

d) The $^3H(d,\alpha)n$ Reaction

The behaviour of tritium in solids must be well understood primarily from the point of view of fusion related materials. Also, it is important to know what changes of tritium profile occur in the Ti–T targets used for production of neutrons by powerful neutron generators.

When measuring the depth distribution of tritium in solids, the β rays from tritium decay do not give very useful information because their emission energies show a continuous spectrum. Therefore a tool of non-destructive depth profiling of tritium at submicron depths, the $T(d,\alpha)n$ reaction was studied in a number of papers [67, 80–82].

In [67] a deuterium beam strikes a target tilted 40° to the beam direction, and the energy distribution of outgoing α particles with energy < 3.5 MeV (Q value is +17.6 MeV) is measured at the angle $\theta = 50°$ (see Fig. 5.9). A 3 μm Al filter is used to stop elastically scattered low energy deuterons and the outgoing ^3He particles produced by the $D(d,^3He)n$ reaction between the incident deuterons and the deuterons formerly accumulated in the target during beam bombardment. The technique is tritium selective. The incident energy of deuterons $E_d = 0.25$ MeV ($d\sigma/d\Omega$ is 120 mb/sr) is higher than the resonance energy (0.11 MeV) for the $T(d,\alpha)n$ reaction, but low enough not to cause considerable reaction yields from other light nuclei in the target ($^3He, ^6Li, ^7Li$).

^3He as a decay product of tritium gives rise to the $^3He(d,\alpha)p$ reaction. The α particles from this reaction are close in energy to those from the $T(d,\alpha)n$ reaction and contribute to the same α-particle spectrum. The cross-sections of these reactions are comparable at energies above 400 keV, below which that of the $^3He(d,p)\,^4He$ reaction decreases with decreasing energy in contrast to the $T(d,\alpha)n$ reaction.

Atom ratios of ^3He to T for the T–Ti targets can be inferred from the count ratios of the proton to α-particle peaks in the particle energy spectra. These are produced by the reactions $T(d, \alpha)n$ and ${}^3He(d, p){}^4He$, which are detected with a surface barrier device placed at $90°$ with respect to the incident beam [82]. The neutrons from the $T(d, n){}^4He$ reaction could induce secondary reactions in the detector. These well-known (n, α_i) and (n, p_i) reactions of silicon could be used as the energy calibration of energy spectra.

The technique allows the determination of tritium concentration within a depth of $\sim 1\,\mu m$ from the surface with sensitivity better than about 10^{15} T at./cm². Sensitivity is comparable to, or higher than, that of the $T(p, n){}^3He$ reaction method. The highest sensitivity is expected for low probing energy, near the resonance at $110\,keV$, where the influence of the main background reactions of the $D(d, p)T$ and ${}^3He(d, p){}^4He$ is negligible. The depth resolution, as calculated by *Okuda* et al. [81] for titanium, varies from $0.04\,\mu m$ at the surface to $0.4\,\mu m$ at a depth of $1.4\,\mu m$.

5.3 Hydrogen Depth Profiling by Elastic Recoil Detection

In the case of elastic scattering, the relationships of kinematics look more simple. Indeed for $Q = 0, m_1 = m_3$ and $m_2 = m_4$, instead of (2.8) we have

$$E_{3,4} = kE_1 \; , \tag{5.5}$$

where k is the kinematic factor given by

$$k = \frac{m_1^2}{(m_1 + m_2)^2} \left[\cos\theta \pm \left(m_2^2/m_1^2 - \sin^2\theta \right)^{1/2} \right]^2$$

for a scattered particle (3) and

$$k = 4m_1 m_2 \cos^2\theta/(m_1 + m_2)^2$$

for the recoil nucleus (4).

Analysis by detecting the elastically scattered particles, e.g. Rutherford backscattering spectrometry (RBS), is widely applied in the determination of a large number of elements. However, the difficulty of this method in profiling the light elements, not to mention hydrogen, is well known. In recent years an effort has been made to employ a more suitable light element analysis, the nuclear recoil method, which is called the elastic recoil detection analysis (ERDA or ERD) technique.

The principle of ERD (first proposed by *L'Ecuyer* et al. [83]) is similar to that of RBS; however, instead of analysing the incident ions scattered from target atoms, in this case the light nuclei recoiling after being hit by an incident particle are detected. Due to conservation of energy and momentum the recoils are scattered only in the forward direction.

In order to measure the hydrogen (as well as other impurities) depth profile by ERD, the recoiling particles resulting from elastic collisions of the bombarding ions are detected at some fixed forward angle θ. The recoil nucleus, having an energy just after collision defined by (5.5), undergoes further energy loss in exciting from the solid. The recoil energy is, therefore, a function of depth. The depth-energy relationship may be found in the small-energy-loss approximation by substituting (5.5) and (2.5′) into (2.6′, 7′) to give

$$x = (kE_0 - E)/\tilde{S} \tag{5.6}$$

for reflection geometry and

$$x = \left[kE_0 - E + \overline{S}_2 l/\sin(\theta - \psi)\right]/\tilde{S} \tag{5.6′}$$

for transmission geometry.

Here $\tilde{S} = k\overline{S}_1/\sin\psi + \overline{S}_2/\sin(\theta - \psi)$ is a stopping parameter as in (3.8), where $\partial E_4/\partial E_1 = k$. The depth-energy relationship follows (2.13) at a large analysis depth. The concentration profile can be found by (2.3) where $\eta = 1$. Quite often relative measurements are performed using hydrogen standards with uniform hydrogen content, and the concentration is evaluated by (2.4).

The experimental setup usually used for ERD analysis is schematically given in Fig. 5.9. In order to stop the large flux of elastically scattered incident ions it is necessary (excluding the case of the use of proton- and neutron-beams) to mount a stopper foil as an absorber in front of the detector. This causes an additional energy loss $\Delta E_a(E)$. Accordingly, the energy $E(x)$ in (5.6) should be found as

$$E(x) = E_d(x) + \Delta E_a(E) \quad ,$$

where $E_d(x)$ is the energy of detected recoils.

Since the kinematic factor, according to (5.5), shows considerable dependence on the scattering angle (see Chap. 3), the resolution of the method is affected to a large extent, compared to NRA methods, by the geometrical spread ΔE_g. This is defined by the effective detector acceptance angle $\Delta\theta$, which in turn depends on the beam size b, the detector diaphragm d, and the sample-detector distance D. The spread $\Delta\theta$ and corresponding geometrical broadening may be expressed [84] by

$$\Delta\theta = \frac{1}{D}\left[d^2 + b^2 \sin^2(\theta - \psi)/\sin^2\psi\right]^{1/2} \tag{5.7}$$

and

$$\Delta E_g = [(E_0 - E_1)\partial k/\partial\theta + (kE_1 - E)\cot(\theta - \psi)]\,\Delta\theta \quad . \tag{5.8}$$

The first term in (5.8) represents the energy spread caused by different scattering angles; the second one reflects the path-length difference in the outgoing particles.

As has already been mentioned in Chap. 3, the presence of an absorbing foil degrades the resolution of the method. Mylar or Al foils are usually chosen as absorbers, because of their low average nuclear charge, to minimize energy straggling.

Any ions, as well as monochromatic neutrons, may be used for hydrogen profiling by ERD methods. It seems reasonable to subdivide these methods into (1) heavy ion elastic recoil detection; (2) elastic recoil detection with He beams; (3) proton-proton scattering and (4) recoil detection with monoenergetic neutrons.

a) Heavy Ion Elastic Recoil Method

It follows from (5.5) that the recoil nucleus energy has a maximum at $m_1 = m_2$, i.e. accelerated protons seem to be preferable for the determination of hydrogen. However, in this case it is difficult to separate the scattering by hydrogen from the background from protons elastically scattered from the sample matrix and impurities, without applying coincidence techniques. The latter complicates the whole process of measurements, thus leading to the preferred use of heavy ions obtained with Van-de-Graaff accelerators, tandem generators and cyclotrons.

Chernov et al. [85] were among the first to show the possibilities of the nuclear recoil method in hydrogen depth profiling by heavy ions. They considered the physical principles of the method and applied ^{14}N ions accelerated by a cyclotron to 16 MeV to study the behaviour of hydrogen implanted in titanium and its alloys as a function of temperature. They also used this method to estimate the absorption cross-section for the N–H bonds present in silicon nitride films. Reflection geometry is used for those samples thicker than the ion range and glancing-beam incidence to the sample surface is often employed in order to decrease the angle θ, thus increasing the energy of recoils. The depth resolution of the aluminum near the surface region was evaluated as 500 Å at an analysable depth of 3 μm; the detection limit was 10^{14} at.H/cm^2.

Figure 5.12 shows the energy recoil spectrum obtained with a thin Nb–Sn film deposited on a polished sapphire substrate by a 30 MeV sulphur bombardment [87]. One can see here the concentration peaks of oxygen and hydrogen in the layer displayed alongside the peaks from surface hydrocarbon contamination. The detection limit for hydrogen can be estimated to be 100 at.ppm. The depth resolution of 10 nm was obtained via the width of the surface contamination peak, while the maximum analysable depth for hydrogen did not exceed 200 nm.

Since hydrogen is known to produce the lowest energy recoils of all species, to ensure its reliable identification in the energy spectrum one has to vary the energy and the type of accelerated ions as well as the absorber thickness in front of the detector, depending on the presence of impurities and their concentration in a sample. The effect of the absorber is demonstrated by Fig. 5.13, where the recoil nucleus energy is plotted against mass for different ion beams and thicknesses of the mylar absorber [87].

Madiba et al. [88] compared the ERD method with ^{35}Cl bombardment to the resonance method based on the $^1H(^{19}F, \alpha\gamma)^{16}O$ reaction. If the concentration of hydrogen and the integral beam charge are equal, then the high speed of the ERD method ensures the necessary statistical accuracy of the analysis with less beam damage to the sample, as the overall concentration profile is obtained at fixed beam energy. Both methods were estimated to have essentially the same analysis sensitivity (2×10^{-4} at.H/at.M and 3×10^{-4} at.H/at.M, respectively). Figure 5.14 [88] shows the recoil proton spectrum and the deconvolved hydrogen depth distribution obtained with a silicon sample. The energy of the incident ^{35}Cl ions was 35 MeV. The surface peak and the implanted profile are well resolved.

Thus the heavy ion recoil method, characterized by a high depth-resolution of a few hundreds of Angströms and a detection limit of 10^{-3}–10^{-4} at.H/at.M, has the potential for providing hydrogen depth profiling with high rapidity and

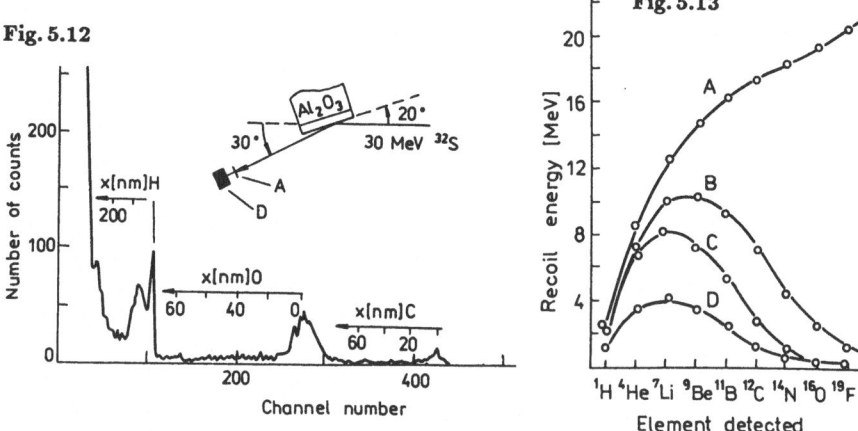

Fig. 5.12. Energy spectrum of recoils from a Nb–Sn layer bombarded by 30 Mev ^{32}S ions. A — absorber, D — detector

Fig. 5.13. Energy of recoils ejected at $\theta = 30°$ by different beams using different absorbers. A: Cl (30 MeV) without mylar, B: Cl (30 MeV) with 10.5 μm mylar, C: Ne (18 MeV) with 8.9 μm mylar, D: Ar (18 MeV) with 7.4 μm mylar

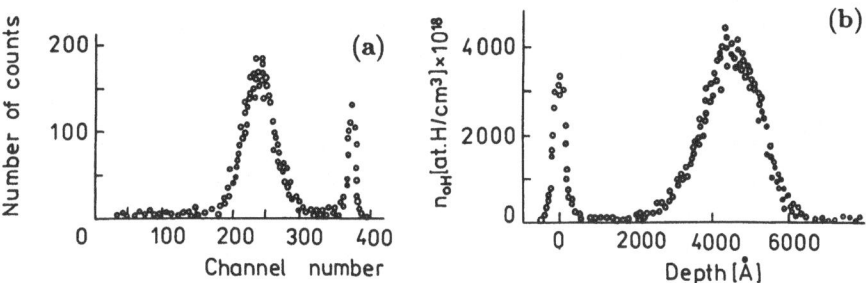

Fig. 5.14. Elastic recoil spectrum from a silicon sample implanted with 35.0 keV hydrogen ions (a); corresponding hydrogen depth profile (b) obtained with 35 MeV ^{35}Cl analysing ions

a maximum profiling depth of up to few microns. It is obvious that the ERD method can be used for depth-profiling any hydrogen isotope; but, in order to simultaneously profile all of the isotopes present, it is necessary to identify the recoils. The latter is not easy to accomplish throughout the entire profiling depth because of the small differences in the mass and energy of different recoils.

b) Hydrogen Depth Profiling by Elastic Recoil Detection with He Beams

The ERD analysis technique using a low energy ^{4}He beam (2–3 MeV) proposed by *Doyle* and *Peercy* [89] is particularly advantageous, as all isotopes H, D and T can be profiled simultaneously with a sensitivity as high as 0.1 at.%. The measurements can be performed using a relatively low-energy accelerator and the target undergoes less damage compared with the use of a higher Z analysing beam.

Turos and *Meyer* [84] discussed in detail the virtues and limitations of this method. The main contributions to the depth resolution have been estimated due to (1) straggling in the absorber foil; (2) geometrical broadening; (3) detector energy resolution and (4) straggling in the sample. The energy straggling in the stopper foil remains the limiting factor. For the choice of the geometry of measurement from the point of view of the maximum analysable depth, they drew the conclusion that the optimum analysis angle is $\theta \approx 30°$, and the tilting angles for incident and outgoing particles should be equal. The variation of incident energy in the range 2–3 MeV does not significantly change the probing depth, because the stopping foil thickness is always adjusted to the range of the incident ions. The maximum probing depth, $x_{max} \approx 500$ nm at $E_0 = 2.5$ MeV, was obtained for a target tilt angle $\psi \approx 16°$; however, the best achievable depth resolution for $x = 100$ nm appears to be $\Delta x = 23$ nm for $\psi = 2°$.

Cheng et al. [90] give the analytical characteristics of the recoil-nuclei method in reflection geometry for the determination of hydrogen in silicon nitride by a glancing beam of 2.1 MeV α particles; the detection limit is 0.1 at.% and the depth resolution is 500 Å at the maximum probing depth of 0.5 μm. They have noted that this method may be conveniently combined with RBS and low-energy NRA techniques in order to gain more information. Using a system for analysing hydrogen and other light elements through the forward α-scattering technique and PIXE (proton induced x ray emission), the total elemental range can be covered [91].

When the ion energy is below the Coulomb barrier, the differential cross-section may be taken as the Rutherford scattering cross-section,

$$\sigma(E) = \frac{1}{\cos^3 \theta} \left[z_1 z_2 e^2 (m_1 + m_2)/2m_2 E \right]^2 , \tag{5.9}$$

where $z_1 e$ and $z_2 e$ are the charges of the projectile and recoil particle, respectively. However, the study of hydrogen concentration profiles with helium ions [92], in which recoil protons are emitted at 0° (transmission geometry), finds the cross-section over the energy range from 1 to 8 MeV to be much larger than Rutherford. This explains the high rapidity of the analysis. The choice of beam energy and sample thickness ensures that the He ions are stopped in the sample, while the recoil protons have sufficient range to reach the detector due to a much lower energy loss. The sensitivity of the technique (*Welunski* et al. [92] worked with samples containing a few percent of hydrogen), is not very high since low Z materials or light element contaminants tend to create a nuclear reaction background that limits the sensitivity. The worsening of depth resolution (compared to other methods) is compensated, however, by the increase in analysable depth. The forward-recoil technique with 4.8 MeV ^{3}He ions is useful for hydrogen depth profiling up to 8 μm of a 25 μm vanadium foil.

Nagata et al. [93] have presented experimental results of depth resolution studies by ERDA of H–Al–H–Si sandwich samples using 3 MeV ^{4}He ions. Typically, a surface resolution of 60 nm was obtained by tilting the target at an angle of 10° with respect to the incident beam. The differential recoil cross-section for H was estimated to be more than double the theoretical Rutherford scattering value, and that for D is greater than 30 times this value near the resonance energy of 2.1 MeV ^{4}He ($\sim$3000 mb/sr at 21.5°). However, the resonance diminishes gradually as the recoil angle increases.

Sawicki [67] has used a beam of ^{4}He projectiles of energy 2 MeV and $\theta = 30°$ for a simultaneous observation of all hydrogen isotopes at a depth 0.1–0.5 μm below the surface in titanium. The recoil cross-sections for H, D and T are respectively 380, 500 and 130 mb/sr, which results in detection sensitivities of 10^{14} at./cm^2 for H and D and about 10^{15} T at./cm^2. However, because of the similar recoil kinematics the recoil energies are close to each other, resulting in the poor separation of isotopes unless particle identification techniques are used. Various methods have been suggested to perform this essential particle-identification function including time-of-flight [94], $\Delta E - E$ analysis with high-energy (25 MeV) α particles from a cyclotron [79], and magnetic analysis [95].

The high precision of the energy analysis provided by a magnetic field was used by *Gozzett* [95] for the simultaneous profiling of hydrogen and other elements present, from H to O, by elastic recoil detection with 3 MeV bombarding ^{4}He ions, resulting in a depth resolution far superior to the methods using usual surface barrier detectors. The magnetic spectrometer technique, equipped with a position sensitive detector, provides a means to achieve a fwhm of about 8.0 nm for a hydrogen surface layer.

c) Hydrogen Depth Profiling by the Proton-Proton Scattering Method

Among a variety of ERD and NRA methods, p-p scattering, using a Van-de-Graaff accelerator [82, 96], tandem generator [97] or cyclotron [98, 99], appears to be the most sensitive technique for hydrogen profiling if two scattered protons are detected in time coincidence. In the opposite case, a major limitation is that high sensitivity can be obtained only if the target is so thin that protons scattered from hydrogen may be kinematically separated from those scattered from the host material. For example, in measurements applied to the determination of tritium loaded in titanium tritide targets [82] the T peak in the 2.5 MeV proton elastic-scattering spectrum is superimposed on the Rutherford-scattering contribution from the thick Mo substrate. Therefore, the detection limit can not exceed $\sim$8 at.%.

Figure 5.15 schematically illustrates the experiment with a 20 MeV proton beam from the 1.5 m cyclotron [99]. Protons are incident on the sample surface at $\psi = 90°$. Since collisions with hydrogen atoms can not result in backscattering, the analysis may be performed by detecting both the scattered and recoiled protons only in transmission geometry. Therefore, the sample placed in the center of the scattering chamber should be in the form of thin foil. For convenience the samples are mounted on a wheel so that they can be changed by remote control. Noting that from the conservation of energy and momentum the angle between the directions of the two emerging protons will be 90° in the laboratory system, they are detected by time-coincidence, using two silicon detectors (D_1 and D_2)

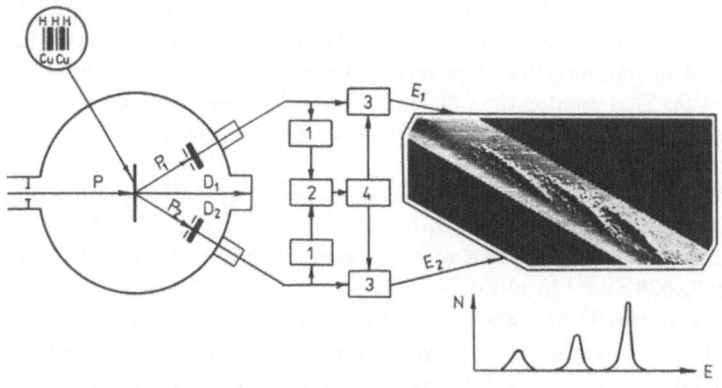

Fig. 5.15. Scattering chamber and associated electronic circuits used for hydrogen profiling by p-p transmission scattering measurements. *1* — time form, *2* — time-to-pulse-height converter, *3* — linear gate, *4* — single-channel-analyser

placed at ±45° to the direction of an incident beam. This renders the detection specific to hydrogen, while all the rest of the beam-matrix interactions appear only as a background of accidental coincidences.

Using an appropriate electronic system, detector signals are fed to the computer-based (MCS-2 consisting of CAMAC equipment and SM-4 computer) two-dimensional multichannel-analyser. The data accumulation program allows one to display the coincidence events as a two-dimensional energy spectrum on the $E_1 - E_2$ axes. The coincidence analysis was accomplished with a time-to-pulse-height converter (TPHC), and its spectra can be displayed simultaneously to control the coincidence resolving time and to set a time acceptance window. Then the region of events relating to hydrogen in the two-dimensional $E_1 - E_2$ spectrum is sampled by the program and projected onto the axis of the summed proton energy to obtain counts as a function of E. The information is then processed to get the hydrogen depth concentration profile. In Fig. 5.15 one can see the graphic display screen, showing the two-dimensional spectrum obtained from a target in the form of a "sandwich" consisting of two 50 μm copper foils alternating with three hydrogen-containing polystyrene films. One can clearly see the hydrogen layers.

Figure 5.16 shows the projection of a two-dimensional spectrum for such a sandwich made of five identical ($\sim 2\,\mathrm{mg/cm^2}$) polystyrene films and four aluminum foils (50 μm thick); the measurements were performed at accelerated proton energies of $E_0 = 17.9\,\mathrm{MeV}$. The number of events on the vertical axis $N(K)$ was obtained by summation along the lines $K_x + K_y = K$ (parallel to hydrogen "ridges") in the field region that is specific to hydrogen. The five peaks in the spectrum correspond to the hydrogen-containing films in the target; their shift relative to one another is mainly determined by the energy losses of incident and scattered protons in the aluminum.

The depth-energy relationship in a good geometry approximation may be found from (2.13), taking into account that, in this case, $\psi = 90°$, $\theta = 45°$,

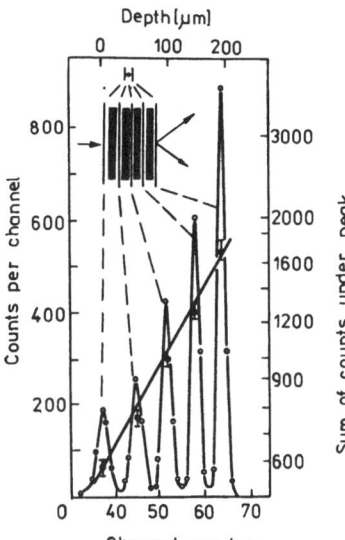

Fig. 5.16. Energy spectrum of p-p coincidences obtained from a H–Al–H–Al–H–Al–H–Al–H sandwich. The straight line illustrates the coincidence count losses due to multiple scattering

$k' = k^\beta$, $k = \cos^2\theta = 0.5$; the depth is read from the surface on which the proton is incident:

$$x = \left\{ \alpha \left[E_0^\beta - (2E)^\beta \right] - l \cdot 2^{\beta+0.5} \right\} \bigg/ (1 - 2^{\beta+0.5}) \ . \tag{5.10}$$

Here $E = (E_1 + E_2)/2$, and l is the sample thickness. The derivative of this expression is positive. This means that, in the measured spectrum, the protons are expected to have the highest energy at $x = l$ and the lowest at $x = 0$.

The maximum thickness of the samples that can be examined is limited by the proton beam energy E_0, the discrimination level in the energy E_{th}, the detector angular aperture $\Delta\theta$ and the range-energy dependence $R = \alpha E^\beta$ for the material examined. Hence,

$$l_{\max} = \alpha \cos(\theta + \Delta\theta) \left\{ \left[E_0 \cos^2(\theta + \Delta\theta) \right]^\beta - E_{th}^\beta \right\} \ . \tag{5.11}$$

At $\Delta\theta = 0.1\,\mathrm{rad}$, $E_0 = 20\,\mathrm{MeV}$ and $E_{th} = 1\,\mathrm{MeV}$, $l_{\max}$ for titanium, copper and palladium is 180, 100 and 90 μm, respectively.

The decrease of the spectrum peak heights in Fig. 5.16 is caused by different coincidence losses due to multiple scattering of particles emitted from different material depths. The straight line indicates that these losses vary exponentially with the channel number, which is proportional to the energy of the detected particle. If we assume such exponential losses (observed also in [97]), they may be taken into account by performing additional measurements with polystyrene thin films of known (standard) hydrogen content, $n_{Hs} = n_{0Hs}l_s$ [cm^{-2}], on the front and back of the sample examined. In this case, one can see two peaks appearing at the edges of the energy spectrum. The total number of counts in these peaks, $N_s(K_1)$ and $N_s(K_2)$, was used to obtain the standard function $n_{Hs}/N_s(K)$, where

$$N_s(K) = \exp\left\{ \ln N_s(K_1) + \frac{(K - K_1)}{(K_2 - K_1)} \left[\ln N_s(K_2) - \ln N_s(K_1) \right] \right\} \quad . \tag{5.12}$$

The hydrogen concentration distribution (at.H/at.M) in the sample analysed was evaluated by a relationship analogous to (2.4),

$$C(x) = \frac{I \, n_{Hs} \, N(K)}{n_{0M} \, N_s(K) \gamma} \frac{dE}{dx} \quad . \tag{5.13}$$

Here I is the ratio of monitor readings in the measurements with the standard and the sample; $N(K)$ is the number of counts in Kth channel of the recorded spectrum after subtraction of the background of accidental coincidences; dE/dx is the derivative of (5.10) (stopping parameter); the energy width of a channel $\gamma = dE/dK$ was determined by calculating the energy of each peak in the spectrum (see Fig. 5.16), using range-energy tables [14, 15].

The possibility of detecting a low hydrogen content is determined by the true-to-accidental ratio of coincidences. *Cohen* et al. [97] showed that a detected limit of 1 ppm may be easily reached using a continuous 17 MeV proton beam from a three-stage Van-de-Graaff accelerator. The use of an accelerator with pulse operation (cyclotron) was shown to lower the sensitivity, raising the threshold to ~10 ppm [99] in spite of the application of two-dimensional analysis of coinciding events in order to decrease the accidental background.

Summing the energies of the coincident pulses from both detectors should exclude the effect of kinematic broadening on the energy resolution, because for each complementary pair of protons we always have $E_1 + E_2 = E$, where E is equal to the energy of an incident beam for an infinitely thin target. Hence, the energy resolution is limited by the beam energy-spread and by the resolution of a detector itself, $\Delta E_D = (\Delta E_{D1}^2 + \Delta E_{D2}^2)^{1/2}$, as well as by the effects of multiple scattering and straggling. The depth resolution of the method, Δx, can be obtained by differentiating (5.10) to give

$$\Delta x = 2^\beta \alpha \beta E^{\beta-1} \Delta E / (2^{\beta+0.5} - 1) \quad . \tag{5.14}$$

It follows that a given energy resolution ΔE leads to a worse depth resolution with the increase of $E(x)$. However, Fig. 5.16 shows that the increase in the distance between the peaks is accompanied by their broadening as a result of straggling and multiple scattering. These two contrary tendencies partially cancel out and make the resolution practically independent of the analysis depth.

Figure 5.17 shows both the dependence $\Delta E(E)$ and the corresponding dependence $\Delta x(x)$ obtained by (5.14) for titanium. The fwhm ΔE were extracted by Gaussian approximation of the peak shapes in Fig. 5.16 and by taking into account their broadening due to the finite thickness of polystyrene film. The depth resolution is estimated as 17, 8 and 7 μm for Ti, Cu and Pd, respectively. The relative resolution (to the maximum analysable depth) in the cases considered is

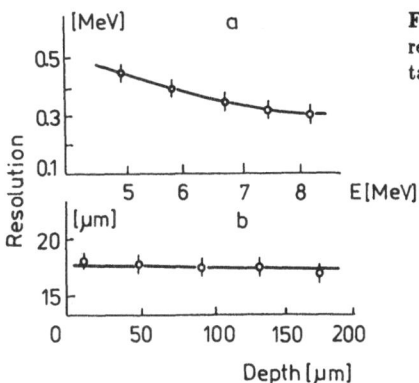

Fig. 5.17. Energy broadening $\Delta E(E)$ (a) and corresponding depth resolution $\Delta x(x)$ for Ti (b) obtained from the energy spectrum in Fig. 5.16

the same and equals $\sim 8\%$. It should be noted that the heavy ion methods which give very high absolute resolution (hundreds of Angströms) in a near-surface region are characterized by practically the same (or even worse) relative resolution as the proton-proton scattering method at the maximum probing depth, since their resolution essentially changes for the worse with depth as shown, e.g., in Fig. 3.2.

As an example Fig. 5.18a [100] shows a two-dimensional spectrum of p-p coincidences (computer display photo) and a corresponding hydrogen depth distribution in a tellurium sample prepared as a plate, 130 μm thick, by electrochemical sedimentation. In this photo one can clearly see the "cross" of accidental coincidences due to intensive proton elastic-scattering on the tellurium matrix. It becomes obvious that there is an advantage in the two-dimensional displaying of detector pulses that allows a decrease of the background of accidentals by selecting only the hydrogen-related region (the central part of the field) for the concentration profile calculation. Alongside the measured hydrogen distribution, the solid line in Fig. 5.18b addresses the accidental background, recalculated into an equivalent concentration profile that determines the detection limit for hydrogen. As can be seen, the detection limit varies from 10^{-5} to 10^{-4} at.H/at.Te depending on depth. The "cross" of accidentals in Fig. 5.18a can be used to calculate the number of background counts under the hydrogen localization region This calculation can be checked by the level of accidental background outside this region. The main sources of error appear to be the statistics of counts in a spectrum and the accuracy of the standard function determination. The figure shows only statistical errors; their increase in distribution from right to left is explained by the lowering of statistical accuracy as a result of coincidence count losses due to multiple scattering. The hydrogen depth distribution in tellurium is characterized by a practically uniform distribution in the bulk with a near-surface concentration increase. As the width of surface peaks corresponds to the instrumental line-width, it can be seen that surface hydrogen is localized in a layer no thicker than ~ 15 μm. However, its comparatively high concentration ($\sim 2 \times 10^{-2}$ at.H/at.Te) cannot be attributed to the usual surface hydrogen contamination, but is probably connected with the mechanism of tellurium electrodeposition.

(a)

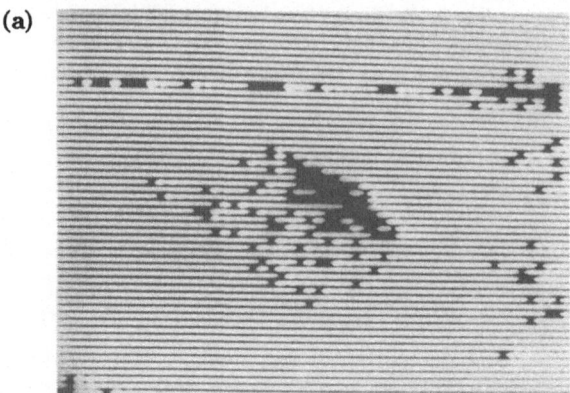

(b)

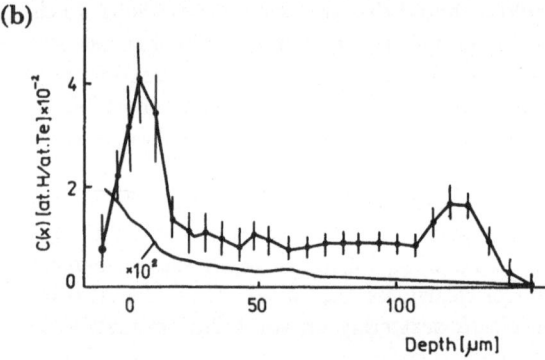

Fig. 5.18. Computer display photo (a) and hydrogen depth profile (b) obtained from a Te sample by the p-p technique. The measurement of x is from the proton-incident surface of the sample. The line $\times 10^2$ represents the accidental background multiplied by 100 times

Attention should be paid to the use of protons instead of heavier ions, as it decreases the influence of the analysing beam on the hydrogen concentration distribution due to defect formation and sample heating. Thus, in the cases considered, if the sample thickness is equal to the maximum probing depth, the 20 MeV incident proton beam loses only ~1.5 MeV of its energy inside the sample as it passes through; intense defect formation occurring at the end of the particle range is also excluded. Besides, the p-p scattering technique has advantages such as high sensitivity and large analysable depth as well as depth-independent resolution. However, this method, applied to samples in the form of thin plates, is not simple to perform and the experimental data are not easy to analyse. Its main disadvantages appear to be the poor absolute depth resolution and the need to make corrections for multiple scattering losses that in turn leads to a variation of sensitivity and analysis accuracy with depth.

The hydrogen depth profiling procedure by proton-proton scattering is rather expensive as it requires the use of large accelerators. However, recently *Willemsen* et al. [96] used a 2.5 MeV Van-de-Graaff accelerator. The applied proton energy-range of 0.5–2.0 MeV makes it necessary to back-etch the area of the sample that is exposed to the beam to a thickness of less than a few microns. Careful adjustment of the experimental conditions gave a proton energy of 1.0 MeV and a scattering angle near 45°, resulting in a detection limit of 0.1 at.% and a depth resolution of 700 Å for a 1600 Å thick Si_3N_4 layer.

d) Hydrogen Depth Profiling by Energy Recoil Detection with Monoenergetic Neutrons

Up to now we have been considering nuclear physics methods of hydrogen profiling by collisions with accelerated charged particles. Monochromatic neutrons have not been used for this purpose for the two following obvious reasons: (1) the low intensity of available neutron sources compared to ion beams; (2) difficulties in the spectrometry of charged reaction products under conditions in which the detector, as well as the sample analysed, is irradiated by a direct neutron beam, resulting in a high background of nuclear reactions with the detector material itself. The development of monochromatic neutron generation techniques along with experimental methods to study (n, charged particle) reactions makes it possible to employ energy recoil analysis by fast-neutron elastic-scattering in hydrogen-isotope depth-profiling [101].

A 150–200 keV deuteron accelerator is used as a neutron source. Monochromatic neutrons with an energy of ∼14 MeV are generated in the nuclear reaction $T(d, n)\,^4He$. The success of the technique is based on the fact that the elastic-scattering cross-section of neutrons on hydrogen isotopes in the case of recoil nucleus emission in the direction of incident neutrons, is higher by one or two orders of magnitude compared to that of the background producing (n, p), (n, d) and (n, t) reactions with the nuclei in the sample matrix, the detectors and environmental materials.

A schematic representation of the experiment is given in Fig. 5.19. A vacuum chamber houses a target wheel and a counter telescope of three surface-barrier silicon detectors. When studying high hydrogen-content materials, the samples and detectors may be in air. The target unit allows remote setting of the sample or standard target between the neutron source and the telescope window. It is also possible to use a diffusion cell with a membrane of the material analysed as a target to study diffusion processes, or naturally any other unit to analyse processes involving hydrogen isotopes.

The neutrons are incident onto a plate of the material to be examined at an angle of $\psi \approx 90°$, thus giving rise to recoil nuclei (protons, deuterons, tritons) in the case of elastic collisions with hydrogen isotope nuclei. The recoils are detected at an angle θ close to 0°. The mean scattering angle $\bar{\theta}$ corresponds to the maximum of the calculated "window" function determined by the neutron source-target-detector geometry. The fwhm of this function $\Delta\theta$ determines the angular resolution of the experimental device. The use of three detectors is dictated by the need to decrease the background by the coincidence technique as well as to identify the type of recoils by the well-known $\Delta E - E$ method. Detector thickness should be chosen according to the hydrogen isotope to be analysed and the analysable depth of interest. The telescope output-signals are displayed

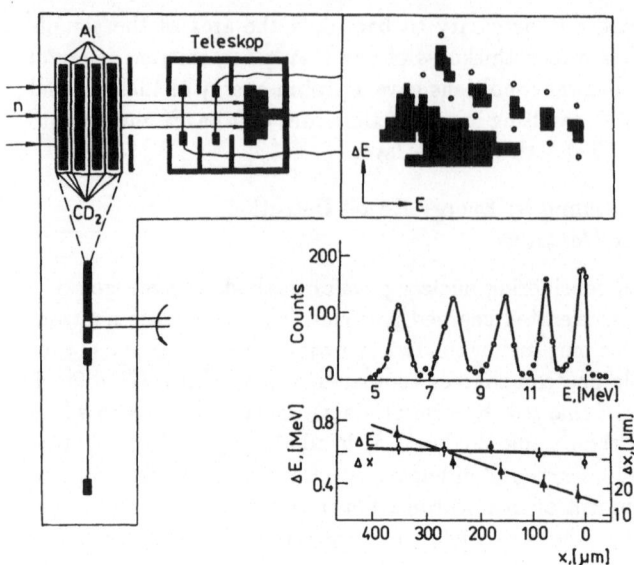

Fig. 5.19. Experimental arrangement of the ERDA technique using monochromatic neutrons. The dots in the two-dimensional spectrum designate the program of deuteron locus sampling

by a computer-based system as a two-dimensional spectrum in the $\Delta E - E$ field, where the events corresponding to different types of particles are localized. Each has its own hyperbola-type locus determined by the known relationship $\Delta E \cdot E \approx \text{const} = mz^2$, where m and z are the mass and charge of the particle detected.

Figure 5.19 shows the energy spectrum obtained from a "sandwich" made of 5 deuterium-substituted polyethylene films ($\sim 2\,\text{mg/cm}^2$) alternated by four aluminum foils (90 μm). In this case the telescope has been formed of two 100 μm transmission detectors, giving two ΔE signals, followed by a 500 μm depletion depth detector providing an energy measurement E. One can clearly see in the spectrum five "bunches" on the deuteron locus corresponding to the deuteropolyethylene films. Background protons are localized on the bottom hyperbola that is characterized by the "turned" shape, as the total thickness of the telescope detectors (700 μm), being sufficient to absorb the maximum deuteron energy (12.9 MeV), is too small to stop protons with an energy more than 10 MeV (the range of protons with an energy of 14 MeV in silicon is $\sim 1.3\,\text{mm}$). This was done intentionally in order to decrease the count rate of the E detector due to neutron-silicon interaction and to improve mass separation in the high-energy part of the spectrum in the case of deuterium profiling.

The program of data processing provides "outlining" of the regions of interest related to different recoils, determination of the projected energy spectra and subsequent calculation of corresponding depth profiles of each hydrogen isotope.

The selectivity of the method consists of the separate detection of protons, deuterons and tritons. The typical mass identification spectrum, shown in Fig. 5.20, illustrates the two-dimensional distribution $\Delta E - E$ obtained by

Fig. 5.20. Two-dimensional spectrum obtained from a Li target bombarded by 14 MeV neutrons. From bottom: proton, deuteron, and triton yields

neutron-induced reactions with a thick target made of a lithium isotope mixture where the high yields of the reactions (n, p), (n, d) and (n, t) make it possible to simulate the presence of all three hydrogen isotopes in the sample. The clear separation of protons, deuterons and tritons in Fig. 5.20 serves as evidence for the possibility of obtaining simultaneous concentration profiles for all hydrogen isotopes. The total detector thickness is ~ 1 mm; hence, the proton energy spectrum is distorted above 12 MeV.

Recoils are emitted from the sample with energy

$$E = kE_0 - \Delta E(x) = \frac{4mE_0}{(m+1)^2} \cos^2 \overline{\theta} - \int\limits_0^{\overline{x}} S_2\left(E(x)\right) dx \ , \tag{5.15}$$

where m is a recoil nuclear mass; $\overline{\theta}$ is the mean angle of its emission $\sim 5°$; $\Delta E(x)$ is the energy loss of the recoil nucleus as it traverses the material from the reaction point at the depth x to its escape from the sample; $\overline{x} = x/\cos\overline{\theta}$ is the average path length, exceeding x (due to the finite angular aperture of the telescope) by less than 1 %. Therefore, let us assume that $\overline{x} = x$. The coordinate may be read either from the entrance or from the exit sample surfaces. The latter is more convenient when the sample thickness is more than the maximum profiling depth.

In the small-energy-loss approximation, the depth-energy relationship is very simple

$$x = (kE_0 - E)/\overline{S} \ . \tag{5.16}$$

In the case of large analysis depth it is sufficient to use only the second equation of the system (2.8). Thus, measuring x from the exit surface and taking $\psi = 90°$, $\theta = 0°$, we also have a simple relationship,

$$x = R(kE_0) - R(E) \approx \alpha \left[(kE_0)^\beta - E^\beta\right] \ . \tag{5.17}$$

In spite of the fact that the measurements are performed with a transmission geometry, the analysis can be carried out essentially without any limitation to the total sample thickness as a result of the high neutron permeability. The maximum analysable depth ($x = x_{max}$), from which the emitted particles still can reach the detector, depends on the material stopping power for the given type of recoils and on the energy threshold E_{th} of their detection,

$$x_{max} = R(kE_0) - R(E_{th}) \ . \tag{5.18}$$

E_{th} is the energy of a particle with range equal in silicon to the full thickness of the first two ΔE detectors, plus the discrimination level in the E detector channel. Figure 5.20 clearly shows the increase of E_{th} due to the cutoff at low energy in the particle spectrum (E_{th}), and the resultant decrease in x_{max} if the isotope mass grows. Since a large part of the neutron energy is transferred to the recoils, the analysable depth may reach a few millimeters, depending on the sample material and the isotope analysed.

In the expression for the reaction yield (2.1) the cross-section $d\sigma(E_0,\theta)/d\Omega$ is constant, as at any depth x neutron-hydrogen collisions occur with the same energy E_0. This results in nearly constant analysis sensitivity throughout the whole depth, and allows one to perform measurements using the total number of counts,

$$N_s = \int_0^{l_s} \left[N(E)\frac{dE}{dx}(E) \right]_s dx = \int_{E_{max}}^{E_{min}} N_s(E)\, dE \ , \tag{5.19}$$

obtained with any hydrogen standard of total hydrogen content $n_{0Hs} \cdot l_s = n_{Hs}$, where l_s is the standard thickness. Then the concentration profile can be expressed (in the units of at.H/at.M) as

$$C(x) = \frac{I\, n_{Hs}\, N(E)}{n_{0M}\, N_s} \frac{dE}{dx} \ . \tag{5.20}$$

For the total content we have

$$C = I N\, n_{Hs}/N_s\, n_{0M}\, l \ . \tag{5.21}$$

Here $N(E)$ is the energy spectrum measured; N is the total number of counts in this spectrum; $l \leq x_{max}$ is the sample thickness analysed; dE/dx is the stopping parameter obtained by differentiating the expressions (5.16) or (5.17), which in our case simply equals $S(E)$, i.e. the stopping power of the sample material for the given type of recoil nuclei with energy E corresponding to the given point in the measured spectrum.

The detection limit for various hydrogen isotopes is determined by the background level in the corresponding regions of the two-dimensional spectrum and may reach 0.1 at.% for deuterium and tritium, but only about 10 at.% for hydrogen. The low sensitivity for hydrogen is explained by the high level of background protons appearing mainly as a result of nuclear reactions with the sample material and the silicon of the first detector. The replacement of the first detector by a proportional gas counter filled with CO_2 and careful choice of the telescope geometry and diaphragm materials makes it possible to decrease considerably the proton background [102] and thus improve the hydrogen sensitivity.

It should be noted that hydrogen depth concentration profiling assumes a uniform hydrogen distribution across the sample area visible to the telescope aperture. In order to increase the reaction yield this area is chosen rather large (~ 1 cm^2); hence, the possible non-uniformity of the lateral hydrogen distribution may result in an additional, undefined error.

In order to determine the depth resolution of this method let us examine again Fig. 5.19, which shows the energy projection of a deuteron locus. The five peaks in the energy spectrum correspond to the overall profiling depth and allow one to calculate the fwhm of each peak ΔE using a Gaussian approximation and the least-squares technique. Thus, one can obtain the depth dependence of the energy resolution, which is shown at the bottom of Fig. 5.19. The width of the peak at the highest energy value is mainly determined by the geometry, $\Delta E_g = 2E \tan \bar\theta \cdot \Delta\theta$ (typically $\bar\theta = 4\text{-}6°$, $\Delta\theta = 4\text{-}8°$), and the additional broadening of the rest of the peaks may be explained by the contribution of both straggling and multiple scattering in the sample material.

Differentiation of (5.17) gives the relationship between energy resolution and depth resolution,

$$\Delta x = \alpha \cdot \beta \cdot E^{\beta-1} \Delta E \ . \tag{5.22}$$

The $\Delta x(x)$ dependence calculated according to (5.22) is also shown in Fig. 5.19 for aluminum. The nature of this dependence indicates that the depth resolution is practically the same throughout the whole profiling depth, as is the case in the p-p scattering method. It is 7–8 % of the maximum probing depth, but may be improved by reducing the geometrical spread ΔE_g. This is possible in the case of a higher intensity neutron source.

Figure 5.21 shows the two-dimensional field of events obtained from a titanium foil 100 μm thick saturated with a mixture of hydrogen and deuterium. It

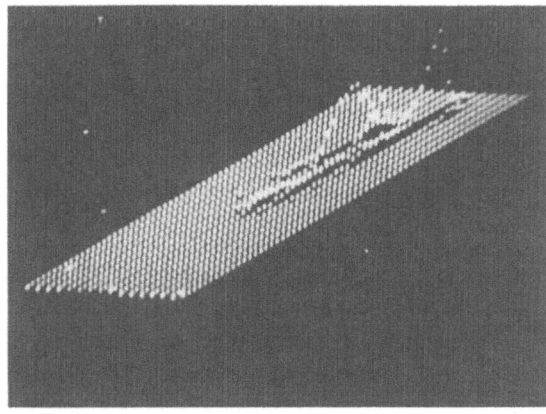

Fig. 5.21. Two-dimensional spectrum (display photo) obtained from a Ti plate containing a mixture of two hydrogen isotopes

may be seen that there is no overlap between the yields caused by different hydrogen isotopes in the spectra. The peak on the ΔE axis which was accumulated during each exposure corresponds to the $^{28}\text{Si}(n, \alpha_0)\,^{25}\text{Mg}$ reaction with silicon in the ΔE detector used as a monitor of the neutron flux. The corresponding concentration profiles are given in Fig. 5.22a and b. Smearing of the edges of the distributions may be explained by the depth resolution of 37 μm for H and 23 μm for D. The lines in Fig. 5.22 result from the convolution of the rectangular distribution of hydrogen and deuterium with the Gaussian resolution function. Their shape indicates a uniform distribution of hydrogen and deuterium throughout the thickness. The experimental distribution in Fig. 5.22a does not at first sight seem to be uniform, as the sample thickness of 100 μm, which is much less than the maximum analysable depth for hydrogen in titanium ($\sim$700 μm), is comparable to the depth resolution of the method. The uniform nature of the hydrogen concentration profile is proved by Fig. 5.22c, where one can see the hydrogen distribution throughout three-such-foils placed together. The smearing of the right edge of the distributions in Fig. 5.22 is probably due to the friable structure of hydrogen-rich plate surfaces. The proton background is represented in Fig. 5.22c by crosses. The deuterium background in this scale is negligibly low.

Along with disadvantages such as low sensitivity for hydrogen and poor absolute resolution, a number of essential advantages of the fast-neutron method should also be mentioned: (1) the large analysable depth along with high neutron-penetrability allows one to investigate the deep regions of the sample, which may be placed in any device for material-hydrogen system study; (2) the possibility of simultaneous concentration depth-profiling of all hydrogen isotopes allows one to study isotopic effects; (3) low intensity irradiation of a sample ($\sim$10^7 cm^{-2}s^{-1}) ensures the absence of sample heating and the minimum effect of radiation damage for the concentration profile of interest; (4) the simplicity of low-energy accelerator design and that of the depth-profile-calculation technique provide an appropriate and cheap method for the determination of hydrogen.

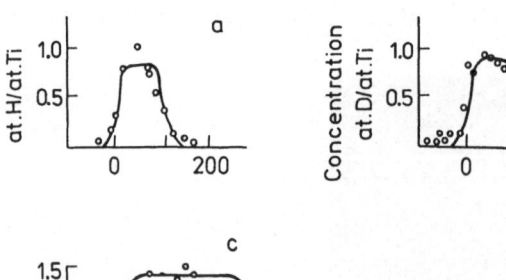

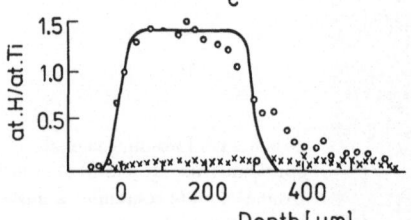

Fig. 5.22. Hydrogen (a,c) and deuterium (b) depth profiles in Ti foils. The proton background is represented by crosses (b)

6.   Hydrogen Diffusion Studies by Means of Depth Profiling

Recently, much emphasis has been placed on the diffusion and trapping of hydrogen isotopes in various metals, both due to their importance in basic physics, and because they have a series of important technological consequences, notably for fusion research and hydrogen embrittlement. A variety of traditional techniques of studying hydrogen diffusion (e.g. the permeation method, the electrochemical method and mechanical relaxation) are rather indirect ways because of the models required for the evaluation of diffusion parameters, i.e. they do not measure the time-variation of the depth-distribution of hydrogen-concentration that determines a diffusion coefficient.

As nuclear physics methods allow one to follow the changes of hydrogen concentration inside the material in a non-destructive way, in principle any of them make possible a direct way of studying the diffusion rate of hydrogen atoms in solids. Indeed, the solution of Fick's diffusion equation $\partial^2 C/\partial x^2 = D^{-1}\partial C/\partial t$ under certain boundary conditions (see examples in [103, 104]) results in an expression for a concentration depth profile $C(x)$ of diffusion species as a function of time t and diffusion coefficient D. Therefore, diffusion data can be obtained by observing the time-dependent hydrogen depth-profiles.

The depth resolution of nuclear physics methods depends on the type of a nuclear reaction used (see Table 3.1) and varies over a wide range. Hence, according to the expression for diffusion length,

$$x = (2Dt)^{1/2} \ , \tag{6.1}$$

these methods allow one to carry out experiments over a wide range of diffusion coefficient values $(10^{-6}$–$10^{-16}\,\mathrm{cm}^2/\mathrm{s})$.

We now consider examples of hydrogen diffusion studies performed by nuclear depth profiling methods, which have only recently become widely used.

6.1   Nuclear Reaction Analysis (NRA) Techniques

When diffusion in a semi-infinite body is studied under the boundary conditions $C = C_0$ for $t = 0$; $C = 0$ for $x = 0$ and $t > 0$; the solution satisfying Fick's laws is

$$C(x,t) = C_0 \operatorname{erf}\left[x/2(Dt)^{1/2}\right] \ , \tag{6.2}$$

where erf is a known error function. These conditions are used in the study of the process of deuterium release after electrocharging of stainless steel by deuterium, performed by *Lewis* and *Farrel* [105]. After charging, the specimen was transferred as quickly as possible to a scattering chamber, where it was bombarded with a 0.45 MeV deuterium ion-beam in order to measure the concentration depth-profile of near-surface deuterium. This was done by energy analysis of protons from the D(d, p) reaction, using a surface-barrier detector. The diffusion coefficient was extracted by fitting the depth profile calculated from (6.2) to the experimental data using the least-squares technique. In spite of the fact that the initial condition $C = C_0$ in (6.2) is not exactly met due to the finite charging time, corrections for this were not found to be important. The temperature during release was assumed to be ambient (298 K). The diffusion coefficient of deuterium in stainless steel corresponding to the minimum in the χ square is $D = (1.4 \pm 0.2) \times 10^{-12}\,\text{cm}^2/\text{s}$, which implies that this technique is a valuable tool in the determination of small diffusion coefficients. It should be noted that the depth profile is sensitive to the diffusion coefficient only in the near-surface region ($x \leq 0.5\,\mu\text{m}$), while the deep region of the spectrum is fully determined by the energy dependence of the $D(d, p)$ cross-section.

The same reaction was used by *Möller* et al. [106] to study deuterium diffusion in palladium at $-160\,^\circ\text{C}$. The loading of palladium with deuterium was done by disconnecting the pumps and the beam tube from the scattering chamber for a short time while admitting deuterium gas. A distinct surface peak is formed during the loading procedure. From this peak some of the deuterium atoms were released into the bulk, allowing extraction of the diffusion coefficient from two subsequent depth profiles, measured after switching on the beam again.

The behaviour of hydrogen implanted into metals is mainly determined by diffusion, trapping and surface recombination. It has been shown by *Möller* et al. [74–76] that all of these effects may be studied with a special dual beam-implantation and analysing device based on a 2.5 MeV Van-de-Graaff accelerator that allows switching from implantation with low-energy D_2^+ ions to analysis by a 790 keV ^{3}He beam. The principle of measurements is shown in Fig. 6.1.

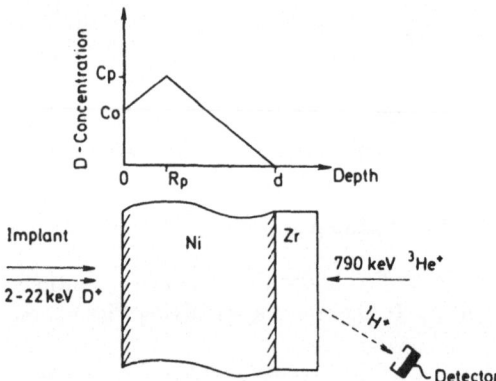

Fig. 6.1. Experiment on deuterium diffusion by means of the $D(^3\text{He}, p)$ ^{4}He reaction. The target holder is rotated by 180° as the implantation and analysing beams come from the same beam port

The initial deuterium concentration was built up by low-energy (2–22 keV) ion implantation into the near-surface region of a stainless steel or nickel membrane 25 μm thick. This made it possible to obtain a deuterium concentration at the upstream side of the membrane exceeding that achievable using the former method of membrane permeation from the gas phase. A thin Ti or Zr layer was evaporated onto the downstream side in order to serve as a trap for the deuterons. The amount of deuterium collected in this getter layer was determined by means of the proton yield detection from the $D(^3\text{He}, p)$ ^{4}He reaction, which served as an integrated deuterium flux through the membrane. If this diffusion flux is plotted as a function of reduced time $\tau = Dt/l^2$, where l is the membrane thickness, then the asymptotic line, according to Fickian kinetics, intersects the abscissa at $\tau = 1/6$ (time lag) from which the diffusion coefficient can be determined.

In the ideal case, i.e. with a fully permeable surface and no traps present, the time lag depends only on the diffusion through the foil, but it appears that the measured curve of permeating fluency and τ vary distinctly with ^{3}He beam intensity. The shape of the curve obtained by frequent analysis is in contradiction to theory. The corresponding deuterium depth profile which was obtained after the permeation experiment shows that the analysing beam produces traps for the diffusing atoms. The damage distribution for 790 keV ^{3}He ions does not differ much from the range distribution. More than 30 % of the total of detected deuterium atoms are to be found around the mean projected range of the ^{3}He analysing beam. Great care must be taken to keep the analysis fluencies so low that they have no measurable influence on the deuterium permeation. On the other hand, the depth-profile measurements allow the determination of binding energy and concentration of bulk traps.

The upper part of Fig. 6.1 shows the steady-state solute concentration distribution for deuterons implanted into a foil. A collector layer at depth d below the surface provides an effective "sink" for the permeating hydrogen at the rear surface assuming a solute concentration $C_d = 0$. The implantation depth corresponds to the projected range R_p. Under steady-state conditions implanted hydrogen will either be re-emitted through the irradiation surface or diffuse through the foil to the opposite side. If the re-emission is purely diffusion-limited, then the solute concentration at the surface is $C_0 = 0$, whereas C_0 increases whenever surface recombination becomes the rate limiting step. Then the surface recombination rate is $K_r C_0^2$, where K_r is the recombination coefficient. In the latter case the permeation rate is correspondingly enhanced. A measurement of the steady-state flux into the gettering layer is then particularly sensitive to even very small changes in the re-emission rate and the recombination coefficient K_r is readily determined.

6.2   Nuclear Resonance Reaction Analysis (NRRA) Techniques

In order to investigate the kinetics and the diffusion mechanism of hydrogen in titanium, zirconium and some alloys, *Dorr, Brauer* et al. [50, 107] used the resonance method based on the ^{1}H(^{15}N, $\alpha\gamma$) ^{12}C reaction. The surfaces of the samples studied were cleaned [107] by annealing at 1300 °C in ultra-high-vacuum (10^{-9} mm of mercury) or by argon-ion sputtering at an accelerating voltage of 4–10 keV. The clean surface was coated with a thin palladium film through which hydrogen loading of the samples was performed by an electrocharging process. Then the samples were placed in the vacuum target chamber to measure hydrogen depth-profiles.

Figure 6.2 shows concentration profiles obtained for three polycrystalline titanium samples after electrocharging for different periods of time. Absorbed hydrogen is concentrated as a hydride layer of varying thickness in the near-

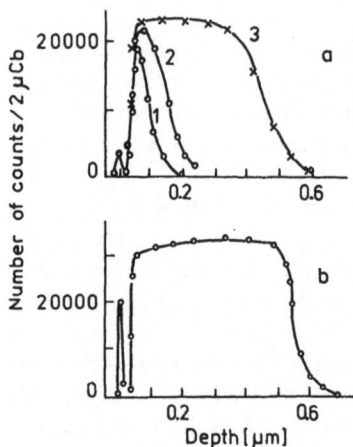

Fig. 6.2. Hydrogen depth profiles in a polycrystalline Ti sample at $t = 6(1)$, $15(2)$, $60(3)$ s (a) and in a monocrystalline one at $t = 100$ s (b) obtained by means of the $^1H(^{15}N, \alpha\gamma)\,^{12}C$ reaction. The peak at $x = 0$ corresponds to hydrogen on Pd-coating

surface region of the metal. When the loading process is over, movement of the hydride boundary into the material ceases. The width of the hydride layer at the half-height of the concentration profile plateau was chosen to be the penetration depth for hydrogen diffusion into the metal, x. The diffusion coefficient was determined by knowing this value and the charging time, t, using the relationship [103]

$$D = x^2/4\alpha^2 t \; , \tag{6.3}$$

where $\alpha = 0.425$ if x is measured in [cm] and D in [cm^2/s]. The value of D for all the samples studied appeared to be an order of magnitude higher than those previously obtained. This may be explained by the more careful cleaning of the sample surface.

As can be seen from Fig. 6.2, the concentration boundary of the hydride layer in polycrystalline samples is smeared, whereas it is sharper in monocrystalline samples. *Brauer* et al. [107] assumed that the reason for this lies in the difference in size and in the random orientation of the grains relative to the surface in polycrystalline material. In its turn, this leads to a supposition about the diffusion mechanism, namely that hydride layer formation proceeds in such a way that each grain should be completely saturated with hydrogen before hydrogen penetrates into a neighbouring, more deeply located grain. Additional arguments for this hypothesis appear to be the high values of D for annealed samples with larger grains compared to those of non-annealed samples, as well as the dependence of D on the orientation of a monocrystalline sample.

The redistribution process of hydrogen implanted into Ta has been investigated in the temperature range of 20–290 K by employing the ^{15}N NRRA technique [108]. Figure 6.3 shows an isothermal annealing sequence of depth profiles at 35 K for hydrogen implanted in Ta at 20 K. In contrast to the results for electrocharged titanium [107], after implantation hydrogen penetrates into the deeper region of Ta. The diffusion coefficient obtained by fitting the experimental diffusion broadening to that calculated by Fickian kinetics in a semi-infinite body

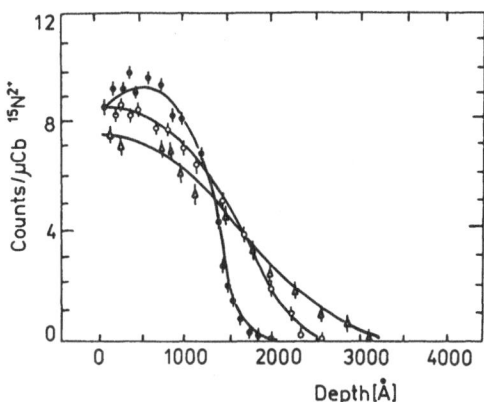

Fig. 6.3. Diffusion broadening of a hydrogen profile in Ta. ($\bullet$) — as implanted (20 K); ($\circ$) — annealing (35 K, 23300 s); $D = 6.6 \times 10^{-16}$ cm^2/s; ($\triangle$) — annealed (35 K, 47150 s), $D = 5.9 \times 10^{-16}$ cm^2/s

with a reflecting surface appears to be $\sim 6 \times 10^{-16}$ cm^2/s at 35 K and shows no dependence on structural properties such as poly- or monocrystallinity, nor on implanted concentrations within 0.2–0.5 %. The diffusion data between 20 and 180 K cannot be described by a classical Boltzmann term, but follow a power law: $D \sim T^n$. This means that in the given temperature interval the quantum-mechanical diffusion of hydrogen in tantalum is observed. At $T > 180$ K a typical Arrhenius-like activation process takes place.

The ^{1}H(^{15}N, $\alpha\gamma$) ^{12}C reaction was also used for a study of hydrogenation in high-purity silicon crystals [109]. The role of hydrogen in silicon has become an area of great interest to semiconductor physicists both from the academic and technological view-points. H$_2$ ions were implanted into high-impurity crystalline silicon at a depth of about 2800 Å. In addition, a defect region was created at a depth of about 670 Å by 70 keV Ar implantation. The study of hydrogen depth profiles before and after annealing shows that hydrogen is rearranged into two defect regions. It has been shown that the damage-free hydrogenation of high-purity crystalline silicon cannot be achieved by using electrolytical or H$_2$ plasma charging. *Chengzhou* et al. [109] have assumed that the incorporation and diffusion of hydrogen in silicon crystals are defect-related in nature.

6.3 Elastic Recoil Detection (ERD) Techniques

Due to the large stopping power of the low-energy D, ^{3}He and ^{15}N probing ions, high depth resolution in the examples of hydrogen diffusion study considered allows the determination of very low values of diffusion coefficients (down to 10^{-16} cm^2/s). However, these methods also have some restrictions related to hydrogen diffusion, the principle ones being:

1) Due to the small analysable depth it is very difficult to determine values of diffusion coefficients higher than 10^{-10} cm^2/s. Under the conditions of fast hydrogen permeation, the system in the sample region being examined will man-

age to relax very quickly compared to the time required for profile measurement. Also, diffusion studies can be carried out only with highly polished surfaces.

2) Since the sample analysed must be placed in a high vacuum, the concentration of hydrogen at a sample surface cannot be maintained.

3) The study of isotopic effects seems to be rather difficult since the peculiarities of the reaction used allow only one hydrogen isotope to be analysed. Usually the detection of another isotope demands an additional run with another projectile.

4) The high-stopping power of the probing ions, particularly in the case of high beam-current values, might cause a radiation-induced distortion of the diffusion process. This is due to the local heating of the sample and to effects like trapping or stress formation which influence the hydrogen permeation in an unpredictable way.

The disadvantages of the methods mentioned (NRA and NRRA) restrict diffusion studies to cases where hydrogen has already been implanted (or absorbed) in a sample. They do not permit investigation of stationary and non-steady-state processes of hydrogen permeation throughout a material from the gas (or gas-liquid) phase under non-varying or varying (if necessary) thermodynamic conditions. They also do not permit investigation of processes with large diffusion coefficients, e.g. in such materials as palladium and its alloys, which are very important for science and technology. However, ERD depth-profiling techniques based on accelerated protons and monochromatic neutrons as the probing particles make it possible to overcome to a great extent the above mentioned disadvantages when used for the study of hydrogen diffusion.

We described [99] the results of a study of hydrogen diffusion in palladium by depth profiling with the proton-proton transmission scattering method (see Sect. 5.3). A palladium membrane $\sim100\,\mu m$ thick was vacuum sealed in a diffusion cell mounted on a target wheel in a scattering chamber, in place of one of the targets (Fig. 6.4). A proton beam collimated to 1 mm in diameter and

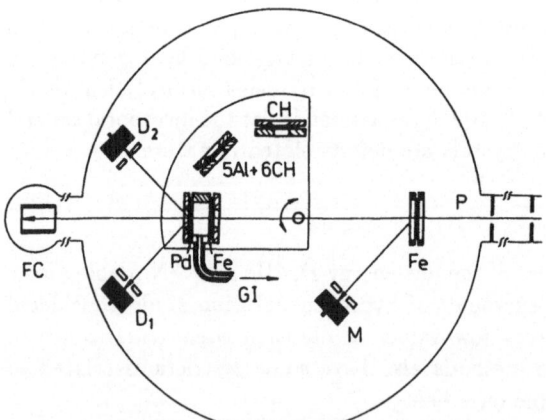

Fig. 6.4. Scattering chamber adapted for hydrogen diffusion study by means of p-p transmission scattering. D and M — are the main and monitoring silicon detectors; Fe — $20\,\mu m$ iron foil; FC — Faraday cup; GI — gas inlet

with energy $E_0 \approx 18\,\text{MeV}$ enters the cell through a thin iron foil. Then it passes through the hydrogen-filled "upstream" part of the cell and a palladium membrane. The system of remote-target-change permitted the placing of additional targets in front of the proton beam. This was necessary in order to calibrate the energy scale and to standardize the measurements. The relative number of protons striking the target was determined by calculating the number of counts in the peak due to their elastic scattering from an iron foil placed just behind the beam collimator. This was necessary since the reading of the Faraday cup depended very much on the thickness of the irradiated target, due to multiple scattering.

As is well known, palladium membranes without a special surface pretreatment allow hydrogen permeation only on heating the membrane above room temperature. A method has been developed for palladium membrane surface activation by coating it with a thin layer of palladium black. This increases the hydrogen permeability tens of times. It is important to note that membranes with activated surfaces separate hydrogen without heating. Thus one can study the diffusion process at low temperatures, where the data are contradictory.

Figure 6.5 shows the two energy-spectra of p-p coincidences obtained with an ordinary palladium membrane (1) and also with a membrane with both surfaces covered by palladium black (2) (activated membrane). In both cases a constant hydrogen pressure of $10^5\,\text{Pa}$ was maintained at the upstream side, while the downstream side was evacuated. Since the measurements were performed at room temperature, conditions were adjusted to provide steady-state permeation of hydrogen from one side of the membrane, which should have an equilibrium concentration corresponding to the β phase of the H–Pd-system, to the other side, where the equilibrium concentration corresponds to the α phase.

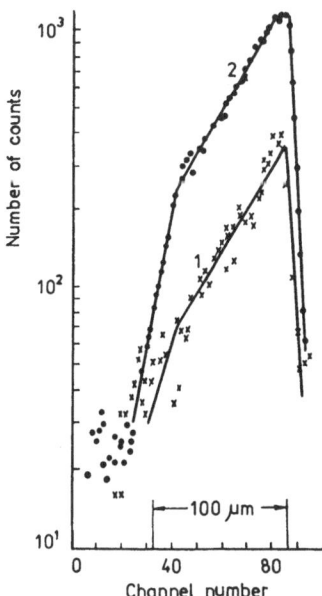

Fig. 6.5. Energy spectrum of p-p coincidences obtained in the process of hydrogen permeating through an ordinary (1) and an activated Pd-membrane (2)

As can be seen from Fig. 6.5, in both cases the energy spectrum is characterized by an exponential shape with resolution-dependent smearing at the edges. One can see in Fig. 5.16 (Sect. 5.3) that such an energy spectrum shape implies a practically uniform hydrogen distribution throughout the whole membrane depth if one takes into account corrections for multiple scattering.

In spite of the large difference in pressure on the membrane surfaces, the hydrogen depth distribution is characterized by having no marked concentration gradient in either membrane. The total amount of hydrogen taken up per Pd atom in the activated membrane bulk (~0.7 at.H/at.Pd) corresponds to the β phase of the hydrogen-palladium system, as can be seen from the pressure-composition isotherm. This means that the permeation process is more limited by the rate of the surface processes at the membrane exit than at its entrance.

A membrane without the catalytic coating (the measurements were performed under the same conditions on the 6th day, following the hydrogen supply to the entrance surface) also contains uniformly distributed hydrogen, but its concentration (~0.03 at.H/at.Pd) corresponds to the α phase of the Pd–H system. This means that in this case there are strong surface limitations for both the upstream and downstream sides of the membrane, which contains a marked amount of hydrogen, although there is no visible permeation flux. The data obtained indicate that the activation of a palladium surface leads to a considerable increase in the hydrogen solubility rate. Also, the supposition can be made that the process of hydrogen permeation through a palladium membrane is more limited by processes at its exit than at its entrance.

Deuterium diffusion in palladium at room temperature was studied using the method of concentration depth-profiling by the elastic scattering of neutrons with energy ~14 MeV [101]. The experimental set-up shown in Fig. 5.19 was changed by using a special diffusion cell made of two chambers separated by a palladium membrane. The cell was connected to a gas-vacuum system that provided vacuum pumping of the cell and gas supply at the necessary pressure, as well as the measurements of the diffusion flux through the membrane.

The process of deuterium sorption by palladium at 10^5 Pa, followed by its desorption when both cell chambers were evacuated, was studied using an activated palladium membrane ~80 μm thick. First, gaseous deuterium was supplied to both membrane sides up to saturation (~0.7 at.D/at.Pd). Then evacuation was carried out. During the sorption-desorption processes the membrane was irradiated to obtain the deuterium concentration profiles, together with the time dependence of the total deuterium content. The dotted lines in Fig. 6.6 crossing the time axis designate the average time measured after deuterium application or degassing, taking into account exposure duration (6 minutes in the process of sorption and 10–30 minutes in the process of desorption). It is seen from Fig. 6.6 that the sorption process differs from that of desorption, both in the total duration, and in the time-dependence of the corresponding concentration profiles.

The solution of the non-stationary Fick's equation when the concentration of deuterium on the membrane surfaces was $C = C_0$ at $t = 0$, taking into account the exposure time τ, was found to be

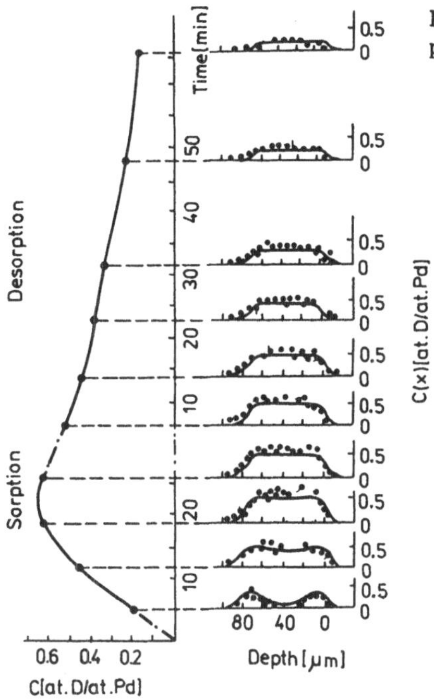

Fig. 6.6. Time sequence of the deuterium depth profile and time-dependence of the total deuterium

$$C(x) = C_0\left\{1 + \frac{2}{\pi D\tau}\sum_{l=1}^{n}\left(\frac{l}{n\pi}\right)^2\left[e^{-t(n\pi/l)^2 D} - e^{-(\tau+t)(n\pi/l)^2 D}\right]\right.$$
$$\left. \times \sin\left(n\pi x/l\right)\frac{\cos(n+\pi)-1}{n}\right\}. \tag{6.4}$$

The solid lines plotted on the sorption profiles were obtained by convolution of the calculated distributions in (6.4) with the resolution function of the instrument and were fitted to the experimental distribution using the diffusion coefficient D. The values $D = 0.5$, 1.15 and 2.15 $\times$ 10^{-8} cm^2/s were obtained by the fitting procedure for the first three sorption profiles. This is much less than the known values for the α as well as for the β phases of the system D–Pd ($\sim$3 $\times$ 10^{-7} [1] and $\sim$10^{-6} cm^2/s^{-1} [110], respectively). As will be shown below, the values of D obtained imply that, under the conditions described, the non-stationary process is complicated by phase transitions in the D–Pd system. Hence, its description by Fickian diffusion kinetics derived for monophase systems is not correct.

The desorption process is much slower than that of sorption. An attempt to describe the desorption profiles in a similar way to the sorption ones, i.e. by solving the corresponding Fick's equation, gave no reasonable results. Moreover, all of the depth profiles observed agree with the supposition of an essentially uniform distribution of deuterium concentration over the membrane thickness. This is illustrated by the solid lines for the desorption profiles, which reflect the convolution of the rectangular concentration distribution by (3.7). From this it follows that deuterium transport through the membrane exit surface proceeds much more slowly, than diffusion in its bulk. Therefore, the deuterium release is not accompanied by the formation of a visible concentration gradient due to fast diffusion relaxation in the bulk of the membrane.

Figure 6.7 represents an experiment [111] for the study of deuterium permeation at room temperature through a 225 μm Pd-membrane with its surface activated by palladium black. At the start of the experiment at $t = 0$, gaseous deuterium was supplied to the upstream side of the membrane at a pressure of 10^5 Pa, maintained constant during the experiment. The downstream side was pumped through the calibration hole to measure the diffusion flux through the membrane. One can see in Fig. 6.7 the time-dependence of the relative values of diffusion flux and deuterium depth profiles in a membrane, obtained in the corresponding time intervals $t = \tau$, where t is the onset of the measurement and τ is its duration for each concentration profile (marked by brackets on the time axis). In this case the coordinate was read from the upstream membrane surface. Since the overall membrane thickness exceeds the maximum analysable depth for deuterium, its first 55 μm occur beyond the spectrometer threshold.

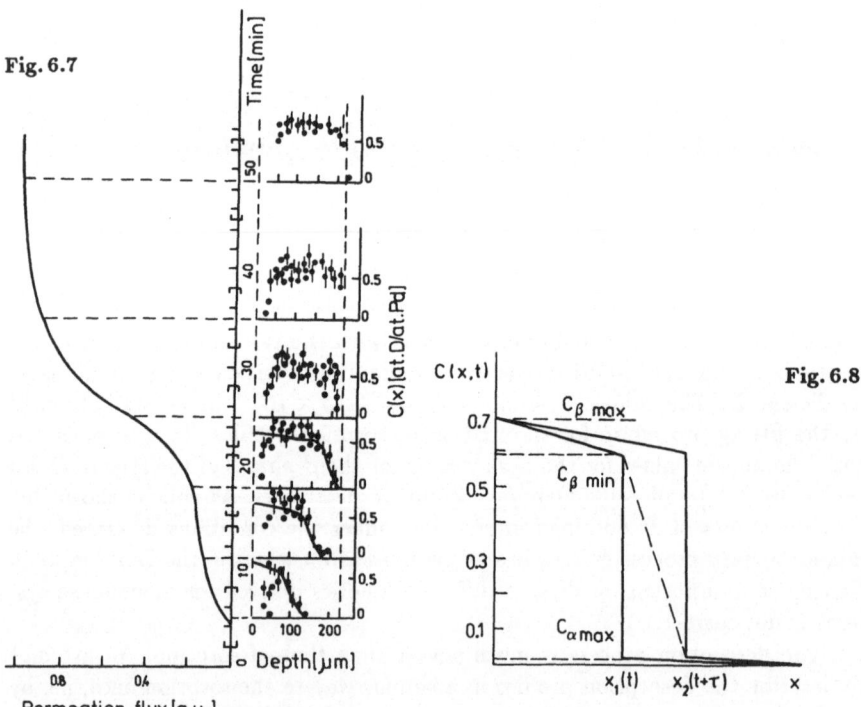

Fig. 6.7. Time-dependence of deuterium depth profile and change of diffusion flux through a Pd-membrane in a non-stationary permeating process

Fig. 6.8. Phase boundary movement inside the Pd-membrane

A gas pressure of 10^5 Pa and room temperature appear to be the conditions corresponding to β-phase formation in the system D–Pd with high deuterium concentration ($\sim$0.7 at.D/at.Pd), as shown by pressure-composition isotherms. Since deuterium was not present in the membrane in the initial state, the β phase cannot be realized without preliminary low-concentration α-phase formation. Thus, a non-stationary permeation process should be accompanied by an α–β phase transition in the D–Pd system. Due to this, the curve for the time-dependence of the flux should not have the characteristic shape of the non-stationary process in the absence of a phase transition (concave instead of convex). The shape of concentration profiles is also not characteristic of the process in a monophase system (convex profile instead of a concave one).

This unusual behaviour of the diffusion flux and the concentration profile shape with time may be explained if one supposes that (i) the disintegrating and forming phases are in quasi-equilibrium, (ii) surface limitations on diffusion are absent on the upstream side and (iii) the phase transition in the system is so fast that the diffusion appears to be the rate-limiting step for the growth of the β phase. If this is the case, one may consider the movement of a flat phase-boundary inside a membrane where the concentration changes by a jump from $C_{\beta min} \approx 0.6$ at.D/at.Pd (the minimum concentration in the β phase) to $C_{\alpha max} \approx 0.01$ at.D/at.Pd (the maximum concentration in α phase) (Fig. 6.8).

We write the continuity equation for the flux through the phase boundary using the approximation of a linear concentration distribution in each phase,

$$\Delta C_{\alpha\beta} \cdot dx/dt = D_\beta \cdot \Delta C_\beta/x - D_\alpha \cdot \Delta C_\alpha/(l - x) \ . \tag{6.5}$$

Analysing the solution of (6.5) one may see that, under the specified experimental conditions, through almost the entire membrane thickness (at least for the first three concentration profiles in Fig. 6.7) the boundary moves according to

$$x = at^{1/2} \ , \tag{6.6}$$

which is true even in the case of the existence of kinetic diffusion-limitations on the exit surface. Here the constant a determines the movement rate of the phase transition boundary in a membrane,

$$dx/dt = a/2t^{1/2} \ , \tag{6.7}$$

and it is connected with the diffusion coefficients in the $\beta(D_\beta)$- and $\alpha(D_\alpha)$-phases by

$$a = [2(D_\beta \Delta C_\beta + D_\alpha C_{\alpha max})/\Delta C_{\alpha\beta}]^{1/2} \ . \tag{6.8}$$

The concentrations C and their differences ΔC in (6.5) and (6.8) correspond to the pressure-composition isotherm of the system D–Pd [1]. They are given in Fig. 6.8.

The spectra were recorded during finite time intervals $\tau \approx 6$ min; hence, the flat front of a phase transition in the distribution measured should be distorted due to its movement during the exposure time (see dotted line in Fig. 6.8). If at the beginning of the exposure the boundary has the coordinate $x_1 = at^{1/2}$ and at the end $x_2 = a(t + \tau)^{1/2}$, then the measured distribution will be

$$C(x,t) \approx (C_{\beta\min}/\tau)(t + \tau - x^2/a^2) \ , \ x_1 \leq x \leq x_2 \ . \tag{6.9}$$

The very low deuterium concentration in the $x > x_2$ (i.e. α phase) depth-interval could not be measured in the experiment due to the short exposure time.

Comparison of the three concentration profiles with (6.9), first transformed according to (3.7), showed (see solid lines in Fig. 6.7) good agreement between the calculated and experimental distributions for the same parameter a found by the fitting procedure. This proves the supposition that deuterium permeation is diffusion-limited and also the fact that there really is movement of the flat α–β phase-transition boundary in the non-stationary process of β-phase formation in the D–Pd system. The value of a found by taking into account the known value of $D_\alpha = 3.7 \times 10^{-7}\,\mathrm{cm}^2/\mathrm{s}$ corresponds to $D_\beta = 1.6 \times 10^{-6}\,\mathrm{cm}^2/\mathrm{s}$, i.e. to the real value. It is of interest to note that the determination of the diffusion coefficient on the basis of the onset time of steady-state permeation, as in the case of monophase systems, leads to the effective value $D < 10^{-7}\,\mathrm{cm}^2/\mathrm{s}$, as in the sorption process described above.

The process studied may be presented in the following way. When gas is supplied to the upstream side of the membrane (the pressure changes from 0 to $10^5\,\mathrm{Pa}$), first the α phase appears on its surface, characterized by a lower equilibrium pressure, then it starts to develop into the bulk. When it reaches the exit surface, a diffusion gas-flux at the downstream side of the membrane appears. The β phase is also immediately nucleated on the entrance surface after the gas supply starts. However, due to the high value of $C_{\alpha\beta} = C_{\beta\min} - C_{\alpha\max}$, the β phase grows more slowly than the α phase, in spite of the fact that $D_\beta > D_\alpha$. The "concavity" of the exit flux curve (see Fig. 6.7) just reflects the kinetics of the β-phase formation. The increase of the deuterium concentration on the membrane exit surface when the β-phase front approaches it leads to the increase of the membrane permeability.

Of course, the movement of the interphase boundary from both sides to the center of a membrane also influences the concentration profiles obtained in the process of sorption (see Fig. 6.6). However, due to the smaller membrane thickness and due to the deuterium permeating simultaneously from both sides, the influence of a phase transition on concentration profiles is not markedly manifested. Nevertheless, it is expressed in the slow-down of the sorption process, resulting nominally in a lower diffusion coefficient.

Thus, the study of diffusion of hydrogen and deuterium in palladium under the conditions described above showed:

1) Surface activation by palladium black allows one to consider hydrogen sorption as a diffusion-limited process, whereas desorption is fully limited by the low-rate processes on the exit surfaces.
2) The α–β phase transition that accompanies the non-stationary process of hydrogen permeation through the palladium membrane is observed as the movement of a flat interphase boundary and leads to the efficient slowing down of both the sorption and the permeation processes.

The above examples of hydrogen diffusion studies based on nuclear physics methods of depth profiling demonstrate the possibilities of this new, direct method that permits one to obtain qualitatively new information, unavailable before, on hydrogen behaviour in solids. This information poses some new questions, such as the following:

— what is the reason for the observed slow-down of the desorption process as compared to hydrogen sorption by palladium, and what is the role of the reverse β–α transition that appears to occur in the course of hydrogen desorption from a membrane saturated to β phase in vacuum?
— what is the character of the isotopic effects that may appear inside a material in the study of the diffusion of an isotope mixture under the conditions of phase transition, and is this true either for one isotope or for all of them?
— would the problem of hysteresis in hydride-like materials be clarified if we studied the pattern of concentration profiles caused by pressure and temperature variations?

These questions need, for their solution, detailed hydrogen diffusion studies employing the above described methods.

7.   Conclusions

The methods of nuclear physics developed in the 1970s as valuable tools in the investigation of hydrogen-material systems soon found applications in the solution of such urgent problems as the behaviour of hydrogen in materials used in nuclear, thermonuclear and hydrogen energetics, the influence of hydrogen on the properties of such materials as amorphous silicon used for the production of cheap solar cells and some microelectronics devices, the role played by hydrogen in superconductive materials, etc.

When suitable radiation sources are used, these methods allow one to determine the total hydrogen content and also to carry out non-destructive studies on the hydrogen concentration distribution in solids, analysing with sensitivities from one tenth to one millionth part of at.H/at.M, depth resolutions from dozens of Angströms to dozens of microns, and probing depths from one to hundreds of microns. Such a variety of methods permits one to solve a great number of problems of a scientific and applied character.

From our point of view, the most potentially useful nuclear physics methods for obtaining qualitatively new information that is beyond the scope of traditional techniques appear to be time-dependent depth-profiling experiments to investigate the diffusion rate of hydrogen-isotope atoms in solids. The results of the direct study of the diffusion process are of great interest, not only from the theoretical point of view; they are very important for economic reasons (reliable and efficient membrane filters for hydrogen, storage and separation of hydrogen isotopes in hydride phases, fusion-related materials, etc.). The range of scientific and technical problems where nuclear physics methods may play a crucial role is not fully exhausted.

Further study of nuclear reactions, the development of accelerators and detection systems and also the wide use of computers will contribute to the improvement of existing analytical techniques and to the creation of new nuclear physics methods for hydrogen determination. This review is intended to serve merely as an introductory guideline to the subject and the authors hope that it will contribute to further implementation of the methods discussed in various fields of science and technology.

References

1   G. Alefeld, J. Völkl (eds.): *Hydrogen in Metals 1,2* (Springer, Berlin, Heidelberg 1978)
2   *Hydrogen in Metals*, Proc., Int. Symp., Z. Phys. Chem., N.F., **143** (1985)
3   *Properties and Applications of Metal Hydrides V*, Proc. Int. Symp., Maubuisson, France, May 25–30, 1986, J. Less-Common Metals **129** (Part 1), **130** (Part 2), **131** (Part 3) (1987)
4   A.V. Shreider: *Hydrogen in Metals* ("Znanie", ser. "Chimija", Moskow, 1979) (In Russian)
5   M.H. Brodsky, M.A. Frisch, J.F. Ziegler, W.A. Landford: Appl. Phys. Lett. **30**, 561 (1977)
6   V.A. Goltzov: J. Mat. Sci. Eng. **49**, 109 (1981)
7   H. Züchner, B. Hüsser: Z. Phys. Chem., N.F., **147**, 35 (1986)
8   J. Bottiger, S.T. Picraux, N. Rud: *Ion Beam Surface Layer Analysis*, Proc. 2nd Int. Conf., Karlsruhe, September 15–19, 1975 (Plenum Press, New York 1976) pp. 811–820
9   J.F. Ziegler and 26 others: Nucl. Instr. and Meth. **149**, 19 (1978)
10   S.T. Pixraux: *Ion Beam Surface Layer Analysis*, Proc. 2nd Int. Conf., Karlsruhe, September 15–19, 1975 (Plenum Press, New York 1976) pp. 527–537
11   P. Peiche, A. Weidinger, P. Ziegler: Z. Phys. Chem., N.F., **143**, 197 (1985)
12   I.N. Beckman: J. Phys. Chim. **4**, 2785 (1980) (In Russian)
13   S.V. Starodubtzev, A.M. Romanov: *Passage of Charge Particles through Matter* (Acad. Nauk Uzbekskoj SSR, Tashkent 1962) (In Russian)
14   N.N. Pucherov, S.V. Romanovski, T.D. Chesnokova: Tables of Stopping Powers and Ranges for Charged Particles with Energy 1–100 MeV ("Naukova Dumka", Kiev, 1977) (In Russian)
15a   H.H. Andersen, J.F. Ziegler: *Hydrogen Stopping Powers and Ranges in all Elements* (Pergamon Press, New York 1977)
15b   J.F. Ziegler: *Stopping Cross Sections for Energetic Ions in all Elements* (Pergamon Press, New York 1980)
16   A. Turos, L. Wielunski, J. Jelinska: Acta Phys. Polon. **A43**, 657 (1973)
17   B. Maurel, G. Amsel, J.P. Nadai: Nucl. Instr. and Meth. **197**, 1 (1982)
18   G. Amsel, B. Maurel: Nucl. Instr. and Meth. **218**, 183 (1983)
19   B. Maurel, G. Amsel: Nucl. Instr. and Meth. **218**, 159 (1983)
20   G. Amsel, C. Cohen, B. Maurel: Nucl. Instr. and Meth. **B14**, 226 (1986)
21   V.T. Tustanovski: *Evaluation of the Accuracy and Sensitivity of the Activation Analysis* ("Atomisdat", Moskow 1976) (In Russian)
22   A.A. Blank, E.L. Grinzide, B.Ja. Kaplan, L.S. Nadezhdina, A.B. Shaevich, B.Ja. Ufa: J. Anal. Chim **30**, 2058 (1975) (In Russian)
23   M.A. Pick, A. Hanson, K.W. Jones, A.N. Goland: Phys. Rev. **B26**, 2900 (1982)
24   E. Segre (ed.): *Experimental Nuclear Physics, Vol. 1* (New York, London 1953)
25   B.G. Skorodumov, Z.P. Kiseleva, V.N. Kadunshkin, I.I. Trinkin: Pribory i Tekhnika Ehksperimenta N3, 50 (1984) (In Russian)
26   W. Möller, M. Hufschmidt, D. Kamke: Nucl. Instr. and Meth. **140**, 157 (1977)
27   A.D. Marwick, P. Sigmund: Nucl. Instr. and Meth. **126**, 317 (1975)
28   D. Schmaus, A. L'Hoir: Nucl. Instr. and Meth. **B2**, 156; 187 (1984)
29   D. Dieumegard, D. Dubreuil, G. Amsel: Nucl. Instr. and Meth. **166**, 431 (1979)

30 M.B. Lewis: Nucl. Instr. and Meth. **190**, 605 (1981)
31 J. Law, D. Hogan: Nucl. Instr. and Meth. **B5**, 67 (1984)
32 P.J.M. Smulders: Nucl. Instr. and Meth. **B14**, 234 (1986)
33 L.P. Starchik, O. Abbosov, A.A. Turinge: Izv. Acad. Nauk Latvijskoj SSR, ser. phys. i tekhn. nauk N2, 10 (1971) (In Russian)
34 S. Hayashi, H. Nagai, M. Aratani, T. Nozaki, M. Yanokura, I. Kohno, O. Kuboi, Y. Yatsurugi: Nucl. Instr. and Meth. **B16**, 377 (1986)
35 G.A. Ivaschenko, L.P. Starchik: in "Nuclear-Physics Methods of Material Analysis" ("Atomizdat", Moskow, 1971) p. 166 (In Russian)
36 V.D. Bang, X.Ch. Thong, Ch.D. Ngiep: JINR Preprint, 18-12235 (1979)
37 J.C. Overley: Nucl. Instr. and Meth. **B24/25**, 1058 (1987)
38 P. Wille, D. Bünemann, H.J. Lahann, H. Mertins: *Neutron Inelastic Scattering*, Proc. Symp., Vienna, October 17–21, 1977 (Int. Atomic Energy Agency, Vienna 1978) pp. 325–338
39 M.R. Hawkesworth, J.P.G. Farr: J. Electroanal. Chem. Interfacial Electrochem. **119**, 49 (1981)
40 J.W. Hanneken, D.R. Franceschetty, R.B. Loftin: Z. Phys. Chem., N.F., **147**, 47 (1986)
41 G. Steyer, J. Peisl: J. Less-Common Metals **130**, 147 (1987)
42 I.V. Mednis: Izv. Acad. Nauk Latvijskoj SSR, ser. phys. i tekhn. nauk N1, 11 (1967) (In Russian)
43 V.A. Muminov, L.V. Navalikhin: *Activation Analysis using Neutron Generators* ("FAN", Tashkent 1979) (In Russian)
44 V.A. Muminov, R.A. Khajdarov, L.V. Navalikhin: J. Phys. Chim. **54**, 2836 (1980) (In Russian)
45 P. Trocellier, Ch. Engelmann: J. Radioanal. and Nucl. Chem. Articles **100**, 117 (1986)
46 H.J. Whitlow, J. Keinonen, M. Hautala, A. Hautojarvi: Nucl. Instr. and Meth. **B5**, 505 (1984)
47 J. Westerberg, L.E. Svensson, E. Karlsson, M.W. Richardson, K. Lundström: Nucl. Instr. and Meth. **B9**, 49 (1985)
48 K.M. Horn, W.A. Lanford, K. Rodbell, P. Ficalora: Nucl. Instr. and Meth. **B26**, 559 (1987)
49 W.A. Lanford, H.P. Trautvetter, J.F. Ziegler, J. Keller: Appl. Phys. Lett. **28**, 566 (1976)
50 R. Dörr, E. Brauer, R. Gruner, F. Rauch, Z. Phys. Chem., N.F., **116**, 1 (1979)
51 W.A. Lanford: Nucl. Instr. and Meth. **149**, 1 (1978)
52 H. Damjantschitsch, M. Weiser, G. Heusser, S. Kalbitzer, H. Mannsperger: Nucl. Instr. and Meth. **218**, 129 (1983)
53 R.E. Benenson, L.C. Feldman, B.G. Bagley: Nucl. Instr. and Meth. **168**, 547 (1980)
54 W.A. Lanford, C. Burman: Appl. Phys. Lett. **41**, 473 (1982)
55 D.S. Leich, T.A. Tombrello: Nucl. Instr. and Meth. **108**. 67 (1973)
56 G.J. Klark, C.W. White, D.D. Alred, B.R. Appleton, F.B. Koch, C.W. Magee: Nucl. Instr. and Meth. **149**, 9 (1978)
57 F. Xiong, F. Rauch, C. Shi, Z. Zhou, R.P. Livi, T.A. Tombrello: Nucl. Instr. and Meth. **B27**, 432 (1987)
58 G.M. Padawer, E.J. Schneid: Trans. Amer. Nucl. Soc. **12**, 493 (1969)
59a E. Ligeon, J.P. Bugeat, A.C. Shami: Nucl. Instr. and Meth. **149**, 99 (1978)
59b E. Ligeon, A. Guivarch, J. Fontenille, M. Le Contellec: Nucl. Instr. and Meth. **168**, 499 (1980)
60 F. Rauch: Nucl. Instr. and Meth. **B10/11**, 746 (1985)
61 W.A. Lanford: Nucl. Instr. and Meth. **B14**, 123 (1986)
62 K. Bethge: Nucl. Instr. and Meth. **B10/11**, 633 (1985)
63 J.A. Leavitt: Nucl. Instr. and Meth. **B24/25**, 717 (1987)
64 E. Yagi, S. Nacamura, T. Kobayashi, F. Cano, K. Watanabe, Y. Fukai, T. Osaka: J. Less-Common Metals **130**, 207 (1987)
65 M. Pilakonta, A. Neskakis, K. Papastaikoudis, A.A. Katsanos, H. Wenze, K.N. Klatt: J. Less-Common Metals **130**, 525 (1987)

66   F.H.P.M. Habraken, R.H.G. Tijhaar, W.F. Van der Weg: J. Appl. Phys. **59**, 447 (1986)

67   J.A. Sawicki: Nucl. Instr. and Meth. **B23**, 521 (1987)

68   C.J. Altstetter, R. Behrisch, J. Bottiger, F. Rohl, B.M.U. Scherser: Nucl. Instr. and Meth. **149**, 59 (1978)

69   D. Dieumegard, D. Dubreuil, G. Amsel: Nucl. Instr. and Meth. **168**, 223 (1980)

70   I.S. Giles, C.G. Wilson: Nucl. Instr. and Meth. **B21**, 72 (1987)

71   F. Besenbacher, S.M. Myers, P. Nordlander, J.K. Nørskov: J. Appl. Phys. **61**, 1788 (1987)

72   R. Ilic, C. Altstetter: Nucl. Instr. and Meth. **185**, 505 (1981)

73   C. Engelmann, J. Bardy: Nucl. Instr. and Meth. **218**, 209 (1983)

74   W. Möller, B.M.U. Scherzer, R. Behrisch: Nucl. Instr. and Meth. **168**, 289 (1980)

75   P. Børgessen, B.M.U. Scherzer, W. Möller: Nucl. Instr. and Meth. **B7/8**, 67 (1985)

76   P. Børgessen, B.M.U. Scherzer, W. Möller: Nucl. Instr. and Meth. **B9**, 33 (1985)

77   S.M. Myers, F. Besenbacher: J. Appl. Phys. **60**, 3499 (1986)

78a J.C. Davis, J.D. Anderson: J. Vac. Sci. Technol. **12**, 358 (1975)

78b J.C. Davis, H.W. Lefevre, C.H. Poppe, D.M. Drake, L.R. Veeser: Nucl. Instr. and Meth. **149**, 41 (1978)

79   L.G. Earwaker, J.B.A. England, D.J. Goldie: Nucl. Instr. and Meth. **B24/25**, 711 (1987)

80   I.Ja. Barit, L.E. Kuzmin, S.A. Makarov: Surface Phys., chim., mech. N11, 49 (1984) (In Russian)

81   S. Okuda, R. Taniguchi, M. Fujishiro: Nucl. Instr. and Meth. **B14**, 304 (1986)

82   G. Zhao, S. Wu, Y. Ren, Z. Zhou, L. Song, J. Wang, L. Shi, L. Kong: Nucl. Instr. and Meth. **B17**, 56 (1986)

83   J. L'Ecuver, C. Brassard, C. Cardinal, J. Chabbal, L. Dêschenes, J.P. Labrie, B.Terrault, J.G. Martel, R. St. Jacques: J. Appl. Phys. **47**, 881 (1976)

84   A. Turos, O. Meyer: Nucl. Instr. and Meth. **B4**, 92 (1984)

85a I.P. Chernov, V.A. Matusevich, V.P. Kosyr: Atomnaja Energija **41**, 51 (1976) (In Russian)

85b I.P. Chernov, Ju.P. Cherdantsev, V.N Shadrin: J. Phys. Chim. **14**, 2831 (1980) (In Russian)

85c I.P. Chernov, V.N. Shadrin, Ju.P. Cherdantsev, V.N. Sulema, L.V. Shramova, T.P. Smirnova, V.I. Belyi: Thin Solid Films **88**, 49 (1982)

86   C. Nölscher, K. Brenner, R. Knauf, W. Schmidt: Nucl. Instr. and Meth. **218**, 116 (1983)

87   C. Moreau, E.L. Knystantas, R.S. Timsit, R. Groleau: Nucl. Instr. and Meth. **218**, 111 (1983)

88   C.C.P. Madiba, J.P.F. Sellschop, H.J. Annegarn, B.R. Appleton: Nucl. Instr. and Meth. **218**, 409 (1983)

89   B.L. Doyle, P.S. Peercy: Appl. Phys. Lett. **34**, 811 (1979)

90   H. Cheng, Z. Zhou, F. Yang, Z. Xu, Y. Ren: Nucl. Instr. and Meth. **218**, 601 (1983)

91   T.A. Cahill, Y. Matsuda, D. Shadoan, R.A. Eldran, B.H. Kusko: Nucl. Instr. and Meth. **B3**, 263 (1984)

92   L.S. Wielunski, R.E. Benenson, W.A. Lanford: Nucl. Instr. and Meth. **218**, 120 (1983)

93   S. Nagata, S. Yamaguchi, Y. Fujino, Y. Hori, N. Sugiyama, K. Kamoda: Nucl. Instr. and Meth. **B6**, 533 (1985)

94   J. L'Ecuer, C. Brassard, C. Cardinal: Nucl. Instr. and Meth. **149**, 271 (1978)

95   C.R. Gozzett: Nucl. Instr. and Meth. **B15**, 481 (1986)

96   M.F.C. Willemsen, A.M.L. Theunissen, A.E.T. Kuiper: Nucl. Instr. and Meth. **B15**, 492 (1986)

97   B.L. Cohen, C.L. Fink, J.H. Degnan: J. Appl. Phys. **43**, 19 (1972)

98   K.P. Artjomov, V.Z. Goldberg, I.P. Petrov, V.P. Rudakov, I.N. Serikov, V.A. Timopheev: Atomnaja Energija **34**, 265 (1973) (In Russian)

99   V.N. Kadushkin, Z.P. Kiseleva, G.A. Radjuk, B.G. Skorodumov, I.I. Trinkin, V.A. Shpiner, P.K. Khabibullaev, V.N. Serebrjakov: Atomnaja Energija **54**, 49 (1983) (In Russian)

100 V.M. Bigelis, B.G. Skorodumov, G.N. Kim, I.I. Trinkin, V.N. Kadushkin, V.G. Ulanov, L.B. Navalikhin, O.A. Abrarov, P.K. Khabibullaev: Elektrochimija **20**, 384 (1984) (In Russian)

101a A.N. Valiev, V.N. Kadushkin, Z.P. Kiseleva, V.N. Serebrjakov, B.G. Skorodumov, A.P. Sokolov, V.A. Spiner, P.K. Khabibullaev, I.O. Jatzevich: Atomnaja Energija **58**, 27 (1985) (In Russian)

101b B.G. Skorodumov: Pribory i Tekhnika Ehksperimenta N5, 63 (1985) (In Russian)

102 K. Ilakovac, L.G. Kuo, M Petravič, I. Šlaus, P. Tomaš: Phys. Rev. Lett. **6**, 356 (1961)

103 J. Crank: *The Mathematics of Diffusion* (Clarendon, Oxford 1975)

104 E. Fromm, E. Gebhardt: *Gase und Kohlenstoff in Metallen* (Springer, Berlin, Heidelberg, New York 1976)

105 M.B. Lewis, K. Farrell: Appl. Phys. Lett. **36**, 819 (1980)

106 W. Möller, M. Hufschmidt, Th. Pfeiffer: Nucl. Instr. and Meth. **149**, 73 (1978)

107 E. Brauer, R. Gruner, F. Rauch: Ber. Bunsenges. Phys. Chem. N4, 341 (1983)

108 M. Weiser, S. Kalbitzer: Z. Phys. Chem., N.F., **143**, 183 (1985)

109 J.I. Chengzhou, S.H.I. Tiansheng, P. Wang: Nucl. Instr. and Meth. **B12**, 486 (1985)

110 F.A. Lewis: *The Palladium Hydrogen System* (Academic, London, New York 1967)

111 A.N. Valiev, V.N. Kadushkin, Z.P. Kiseleva, V.N. Serebrjakov, B.G. Skorodumov, I.O. Jatsevich, P.K. Khabibullaev: Voprosi Atomnoj Nauki i Tekhniki, ser. AVE i T, vyp.2, 42 (1985) (In Russian)

References Added in Proof

Chernov, J.P., V.N. Shadrin: *Hydrogen and Helium Content Analysis by Means of the Nuclear Recoil Method*, ("Energoatomizdat", Moskow 1988) (In Russian)

Chu, Wei-Kan, David T. Wu: Nucl. Instr. and Meth. **B35**, 518 (1988)

Doyle, B.L., D.K. Brice: Nucl. Instr. and Meth. **B35**, 301 (1988)

Laursen, T., M. Leger, J.L. Whitton: J. Nucl. Mat. **158**, 49–56 (1988)

Lee, S.R., B.L. Doyle: Nucl. Instr. and Meth. **B40/41**, 823–827 (1989)

Musket, R.G.: Nucl. Instr. and Meth. **B40/41**, 591 (1989)

Pretorius, R., M. Peisach, S.W. Mayer: Nucl. Instr. and Meth. **B35**, 478 (1988)

Sokhi, R.S., S.B.A. England, G.M. Field: Nucl. Instr. and Meth. **B40/41**, 809–812 (1989)

Yamamoto, T., S. Okuda, M. Fusishiro: J. Nucl. Mat. **160**, 247 (1988)

Subject Index

Springer Tracts in Modern Physics

Within this long established series there are several volumes on themes which are of interest to colleagues dealing with solid state physics and related aspects:

Volumes 91–95, 98, 99, 101, 103, 104, 106, 107, 109–111

Volume 104

I. Pockrand

Surface Enhanced Raman Vibrational Studies at Solid/ Gas Interfaces

1984. IX, 164 pp. 60 figs. ISBN 3-540-13416-6

"… The monograph well summarises the subject, both from an experimental standpoint, and by way of reviewing the many current theories developed to account for the SERS effect at solid/gas interfaces – particularly at silver surfaces. It is well written and illustrated and well produced; the material is logically presented and certainly to be recommended for researchers interested in the subject."

Journal de Physique

Volume 106

J. Kirschner

Polarized Electrons at Surfaces

1985. VIII, 158 pp. 57 figs. ISBN 3-540-15003-X

Springer-Verlag Berlin Heidelberg New York London Paris Tokyo Hong Kong

"… This book will be of general interest to all who seek and find pleasure in the unity of different physical phenomena…"

Optics News

Springer

Research Reports in Physics

The categories of camera-ready manuscripts (e. g., written in TeX; preferably hard plus soft copy) considered for publication in the Research Reports include:
1. Reports of meetings of particular interest that are devoted to a single topic (provided that the camera-ready manuscript is received within four weeks of the meetings's close!)
2. Preliminary drafts of original papers and monographs.
3. Seminar notes on topics of current interest.
4. Reviews of new fields.

Some titles in the subseries of the Research Reports are:

RRP – Astrophysics

M. Lozano, M. I. Gallardo, J. M. Arias (Eds.)

Nuclear Astrophysics

1989. VIII, 355 pp. 126 figs.
ISBN 3-540-50751-5

RRP – Atoms and Molecules

H. Eschrig

Optimized LCAO-Method

and the Electronic Structure of Extended Systems

1989. 221 pp. 80 figs.
ISBN 3-540-50740-X

RRP – Condensed Matter

M. A. Savchenko, A. V. Stefanovich

Fluctuational Superconductivity of Magnetic Systems

1989. Approx. 180 pp. ISBN 3-540-50561-X

RRP – Ecology

W. Wolff, C.-J. Soeder, F. R. Drepper (Eds.)

Ecodynamics – Contributions to Theoretical Ecology

1988. XI, 351 pp. 116 figs.
ISBN 3-540-50116-9

RRP – Nonlinear Dynamics

J. Engelbrecht (Ed.)

Nonlinear Waves in Active Media

1989. Approx. 300 pp. 85 figs.
ISBN 3-540-51190-3

A. V. Goponov-Grekhov, M. I. Rabinovich, J. Engelbrecht (Eds.)

Nonlinear Waves 1 – Dynamics and Evolution

1989. XII, 238 pp. 91 figs.
ISBN 3-540-50562-8

Nonlinear Waves 2 – Dynamics and Evolution

1989. XII, 176 pp. 44 figs.
ISBN 3-540-50654-3

RRP – Nuclei and Particles

J. Eberth, R. A. Meyer, K. Sistemich (Eds.)

Nuclear Structure of the Zirconium Region

1988. XII, 424 pp. 232 figs.
ISBN 3-540-50120-7

Springer-Verlag Berlin
Heidelberg New York London
Paris Tokyo Hong Kong